Immortality Made Ea$y

EVERYTHING YOU ALWAYS WANTED TO KNOW ABOUT WILLS, TRUSTS, PROBATE, MARRIAGE AND FOUNDATIONS*

By

Paul Rampell, Esq.

Attorney-at-Large

Foreword

By

Donald J. Trump

*but didn't want to pay $500 to some lawyer to find out.

Library of Congress Catalog Card Number 94-12045

ISBN 0-9656791-01

Printed in the United States of America

Parthenon Books, Inc.
666 Fifth Avenue
New York, N.Y. 10103

To my son, Palmer

ACKNOWLEDGMENTS

I wish to express my gratitude for the assistance and alacrity of Phyllis Nola and Patricia Torres who have helped with every aspect of this book.

For their friendship and support, I wish to thank Fred and Betty Brenner, Jo Ann Engelhardt, Kathleen Etherington, Barbara and Jack Evans, Norma I. Foerderer, Bruce P. Gendelman, Joachim Gottschalk, Paul C. Hedrick, Blake Hurt, Frances, Carol and Marjorie Levin, Joseph and Marie Levin, Kathy Merlin, Bob and Mary Montgomery, John A. Moran, Hubert G. Phipps, Richard Rampell, Gregory J. Shibley, Arthur and Belle Silverman, Donald J. Trump and William D. Zabel.

I wish to thank my wife, Rita, and my son, Palmer, for their love.

TABLE OF CONTENTS

After signing his will, Ishmael
said, “I felt all the easier:
a stone was rolled away from
my heart . . . I survived myself.”

—*Moby Dick*
Herman Melville

Foreword

Immortality Made Easy is without question a unique, forthright and brilliant writing.

While some readers may laugh and think of it as comedy, it is really far more than that — but in the guise of comedic prose. In actuality it is a very serious work that should be thoroughly digested by any of the many individuals who fall within the category of needing estate planning, advice and counsel.

I congratulate Paul Rampell on doing successfully what so many authors have failed to do. He has made a complex subject simple and understandable. It is a great privilege for me to write this foreword to this primer.

—Donald J. Trump

CHAPTER ONE

INTRODUCTORY QUESTIONS AND SMART-ALECK ANSWERS

What is this book about?

All of your legal and practical questions will be answered about what to do with your things when you die. Wills and other important family documents will be examined. A Last Will and Testament can make you immortal.

For every good word in your will, you add time to your afterlife.

Why should I read it? Why should I plan to push up daisies?

You spend time planning vacations, preparing for a holiday party or saving for your retirement years. If these merit a little planning, surely you should spend a little time planning for your family, friends or favorite charity when the Grim Reaper gets you.

To address one's death involves an analysis of who you are and what exactly you value. Since you probably are an interesting person, the analysis will be interesting. Consciously or unconsciously, you live by a set of priorities. You attach importance, according to some ranking, to your spouse, children, grandchildren, friends, charities, community, country, business or other phenomenon in your life. When you make plans for the objects of your affection and reduce these plans to writing, in a will or trust or similar document, there's no kidding around about the priorities by which you live and wish to die.

Toni Morrison, in her acceptance speech for the 1993 Nobel Prize for Literature, said, "We die. That may be the meaning of life."

In other words, writing a will can give instant meaning to my life?

Yes.

What's the difference if I do estate planning or don't?

If you plan your estate, you will control certain destinies after your death. If you do not plan your estate, judges, legislatures and various family members may make choices that you failed to make.

For a first example, if you die without a will in Florida, the state legislature dictates in certain laws that about one-half of your money must pass to your spouse and one-half, subject to death taxes, must pass directly to your children. An individual, possibly a total stranger to you and your family, or an institution, such as a bank or trust company, may be selected by a judge to do the dividing and distributing. If you die with a will, you decide who gets what and who does the distributing.

For a second example, if you leave a high-rolling son an inheritance of $500,000 outright, he may spend it all within months after your death and live the rest of his life as a pauper. If instead you create a trust for this child, and he receives monthly payments from the trust, his whole life may be more stable and happy.

"If a man dies and leaves his estate in an uncertain condition, the lawyers become his heirs," said Edgar Watson

Howe. It is true that dying without a will can create significantly greater fees for lawyers.

I don't want to die. I'm afraid to die. Do I have to think about my death?

No one wants to think about his or her death. I assure you that this book will in no way advance the date of your death.

Many of us think we will live forever and under this delusion we defer thoughts about death. Even the most optimistic minister, priest or rabbi will confirm that one day you will die. Let's not argue about this point. By confronting your demise, you can look death in the eyes and defy it. If you have any fears about death, they will disappear by the time you finish reading. After we defeat death, we can enjoy life more and forget about life's conclusion. Wills are your last words. Write yours and make them count.

In other words, if I plan, write and sign my will, you are claiming that I can overcome fears of death?

Yes.

You make wills sound like personal diaries or letters. Are there similarities?

Yes; an excellent comparison. When you write a diary, you put down your most sincere thoughts. It can be psychotherapeutic. The same applies to writing a will. When you write a letter, you can communicate important personal messages to the recipient, whether it's a family member, friend or organization. The same applies to will writing.

Will writing can be cathartic, self-illuminating and — with the right lawyer — very gratifying.

Isn't estate planning a deadly dull subject?

Ha, Ha, Ha.

The question-and-answer format of this book and the irreverent approach of the author are going to excite you. If you come across a question and you already know the answer, or are not interested in the subject, just skip the question and move on to the next one. I won't be offended.

What does marriage have to do with death and taxes?

First, marriage may produce children. Children are very important in estate planning. Second, marriage can reduce death taxes significantly. Finally, some people believe death begins with marriage.

I have a sexual partner but we are not married. My girlfriend and I live together. Will my death taxes change if we get married?

You immoral thing, you. I have never heard of such a situation.

A marriage or the lack of a marriage can have a huge impact on death taxes. One can give his or her spouse an unlimited amount of money without paying a penny in death taxes.

If your sexual partner is married to another person, that marriage will not benefit you at your death any more than it benefits you now. It sounds as if you may live in

Beverly Hills or Palm Beach, the twin capitals of decadence and high death taxes.

Isn't it true that the only two things in life that you can't avoid are death and taxes?

You are paraphrasing Mark Twain, noble reader, who supposedly said that death and taxes are unavoidable. Mr. Twain probably was speaking of income taxes or sales taxes. He did not specifically address death taxes.

Can I avoid death taxes?

You cannot avoid death and taxes, but you can avoid death taxes. If Mark Twain were alive today, with this book now in existence, he would have to qualify his comment.

By the simplest estate planning, described in this how-to book, you can save hundreds of thousands of dollars in taxes when you become worm food. Above $625,000 (increasing gradually to $1,000,000 in 2006), the I.R.S. assesses estate taxes at rates that range from about 37 percent to 60 percent. Obviously these rates are higher than income tax brackets. Read on and reduce your estate-tax bill.

Who the hell are you to write this book?

I am the official attorney of the Olympics! I am an attorney trying to explain the rules of the game of life and death. I have advised a lot of people about family matters, wills, trusts, marriage and related subjects since I received my certificate from correspondence law school. Actually, I did attend Princeton, where I learned to read and write and act irreverently. My clients include many multimillionaires, so my advice must be worth something.

How much does estate planning cost?

This depends upon the complexity of your plans and the price of your lawyer. Wills have been prepared for as little as $100 by some lawyers and as much as $100,000 by other lawyers; it depends upon your circumstances and their reputations and billing practices. This book should save you legal fees because you will understand all of the legal terminology you need and legal options available to you before you even visit an attorney.

Shop legal fees, since they vary wildly.

If I don't read this book, will I go to Hell?

Yes.

If I do read this book, will I go to Heaven?

Yes; I personally guarantee it (assuming you haven't committed too many sins). You will go to Heaven and become a member of The Grateful Dead.

Do you think this book will be made into a movie?

Sure. It will join *Dead Man Walking, Dead Men Don't Wear Plaid, Death Becomes Her, Death Before Dishonor, Death Hunt, Death in Venice, The Death Kiss, Death of a Centerfold, Death of a Scoundrel, Death on the Nile, Death Race 2000, Death Rides a Horse, The Death Squad, Death Wish, Death Wish II, Death Wish III, Death*

Sport, Deathstalker and Deathtrap.

Death is a popular subject in Hollywood. People there just don't like to "do" death. They would rather "do" lunch.

Woody Allen says he's not afraid of death; he just doesn't want to be there when it happens to him. He also says that he doesn't believe in an afterlife, but is bringing an extra pair of underwear just in case.

Is this a serious book?

Deadly serious.

Tell me what you're going to teach me.

It'll be like a romance novel. First, you need to learn about men and women. I write about wills in Chapter Two, like men, and trusts in Chapter Three, like women. Chapter Four addresses executors, trustees and lawyers; these are like matchmakers that you must choose wisely. The probating or carrying out of a Last Will and Testament, like procreation, is discussed in Chapter Five. Chapter Six covers common problems and practical solutions — like planned parenting. Chapter Seven addresses, respectfully, the almighty I.R.S. — like safe sex. Chapters Eight, Nine and Ten enable you to achieve immortality, like eternal love, through thoughtful planning, charitable trusts and foundations.

I provide you with checklists and forms. If you complete these, you are eligible to receive a Guaranty of Immortality from me.

What if I don't get anything out of this book?

You at least will have a legal claim against me. Therefore, I must make this warning. If you rely on anything I write herein to your detriment, I was only kidding when I wrote it.

Don't you just hate it when lawyers write "herein"? If attorneys did not use the words "herein," "therefore," "above described," "aforementioned" and similarly stuffy terms, they would have nothing to sell.

How do you make immortality easy?

By explaining the use of wills, trusts, probate, marriage and foundations as the means to create an afterlife. This is all secular, so please forgive the absence of relgious tips.

Notes from the Underground was Dostoyevsky's will and it's still influential.

He left everything to you on one condition: You must remarry. He wants one person, your next husband, to mourn his death.

CHECKLIST 1-A:

Things Every Dead Person Should Have Done

- Write a list of all of your assets and liabilities. See "Checklist 1-B: Covering Your Assets."
- Write a list of the names and addresses of all of the individuals and organizations you wish to include in your will.
- Check the title to all of your assets to make sure each asset will be transferred as you intend (to a joint owner, to children pursuant to your will, etc.).
- Consult with a lawyer and an accountant to make sure that your will and/or living trust are in order. Check out your lawyer's and accountant's credentials.
- Write an "ethical will," a love letter, to your spouse, girlfriend, boyfriend, children, grandchildren or friends, or all of the above, and express your affection and best wishes for them.
- Talk to your family and friends about your plans.
- Make plans for health care decisions, in case you start to lose your marbles.
- Make plans for decision-making by those you trust in case you turn into a vegetable.
- Write down funeral instructions. See "Checklist 1-C: Getting Ready for Your Own Funeral."
- Think about using your money to instill your values in your family and further your ideals.

CHECKLIST 1-B:

Covering Your Assets

If you complete the following questionnaire, you should cover all of your assets so your heirs don't have to go hunting for buried treasure to find them. Unless you are buried with your assets.

1. What real estate do you own?
   A. Residential
   B. Commercial
2. What checking, savings and money market accounts do you maintain? List banks, addresses, account numbers and contact persons.
3. What stocks and bonds do you own? List stockbrokers and brokerage houses, addresses and account numbers.
4. Describe all life insurance policies and annuities you maintain.
5. Does anyone owe you any money? Describe debts owed to you and include names, notes, mortgages and collateral.
6. Do you have an individual retirement account, pension, profit-sharing plan or other retirement program? If so, describe.
7. Make an inventory of all of your tangible personal property:
   A. Automobiles?
   B. Art work?
   C. Jewelry?
   D. Furnishings?
   E. Boats?
   F. Stamp collections, book collections or other collections?
   G. Junk?
8. Describe any partnership interests, interests in closely held or small businesses or joint ventures.

CHECKLIST 1-C:

Getting Ready for Your Own Funeral

"I haven't got a thing to wear" may be your first thought when preparing for the one party that is thrown for you that you cannot enjoy.

Consider the options about your funeral service, the treatment of your corpse and related arrangements.

- Do you want to be cremated? If yes, where do you want your ashes to go?

- Do you want a funeral service? If yes, who do you want to speak at it? Is there anyone you do not want to speak? Write out any special songs, prayers, poems, long novels or other recitations you would like at the service. You may wish to write out two lists: one list of people you want invited, another list of people you do not want invited and even barred at the door.

- Do you want an obituary published? If yes, in what newspaper? Who would you like to write it? If you wish to write it, get going, because you may check out at any time.

- Do you own a cemetery plot? If you do not, you may wish to buy one. The land is cheaper for the living than for the dead.

- What do you want chiseled on your tombstone as an epitaph? One of my dear friends, who used to be an overactive municipal politician, has written on his: "Gone to another meeting."

CHAPTER TWO

WILLS MADE SIMPLE

Who wrote the meanest will ever and how was it so mean?

According to William D. Zabel, Esq., one of the most vindictive wills ever written was by a German who lived in Munich. It stipulated that a wake be held in an upper story of his residence. When his relatives gathered around his coffin, the floor collapsed and most of the mourners were killed. It was discovered later that shortly before his death, the man had sawed through the supporting beams.

What exactly is a "will"?

A will is (a) the written expression of a person's wishes as to the distribution of his or her property to take effect after his or her death and (b) signed in accordance with the formalities of law.

What might a very simple will say?

A will could be this simple and still be valid:

Will

I give everything I own to my spouse and my spouse shall act as my executor.

Signature for Witness No. 1	Will-maker's Signature
Signature for Witness No. 2	Signature of Notary Public Commission Information and Stamp:

Believe it or not, the above would be a valid, but very inferior, Last Will and Testament. The will fails to address the problem of the spouse dying before the will-maker.

Is an oral or verbal will valid?

No. Can you hear a dead man talking? Voices from séances will not do.

Can I write my own will?

Sure. I recommend this as weakly as I recommend that you perform surgery on yourself.

Jacob M. Brande recites an anonymous poem for lawyers in the *Lifetime Speaker's Encyclopedia*:

When a festive occasion our spirit unbends,
We should never forget the profession's best friends.
So we'll pass round the wine
And a light bumper fill
To the jolly testator who makes his own will.

Is a will written in my own handwriting valid?

Yes, if it is signed in accordance with the law where you live. A handwritten will is referred to as a holographic will.

What formalities of law must be followed in signing a will?

The legal requirements change from state to state, but in most places two adults, preferably independent, and a notary public should witness the will at the same time and in the same room as it is being executed and signed by the will-maker, and the witnessess should sign their names on the will.

If a will is so simple, why do families get into lawsuits over them?

Even though the basic requirements for execution of wills are straightforward, wills may be contested if they were signed under duress, coercion or undue influence.

What is a "will contest"?

An olympic event. Contestants demonstrate their power of will by showing who can hold their breath the longest.

Actually, a will contest is a challenge to the validity of a will for duress, coercion or undue influence or if the will was not signed in accordance with basic requirements.

What is "duress"?

Duress is a state of mind in which a person is not acting under his or her own volition. If a wife was starving her invalid husband until he signed a will leaving her everything, his will would be signed under duress.

What is "coercion"?

Coercion is the situation where one person forces another to do something against his or her desires. If a husband held a gun to his wife's head until she signed a will contrary to her wishes, this will would be coerced. If the wife refused to sign and died, could the husband inherit anything? Stay tuned.

What is "undue influence"?

Undue influence is the effect one person has on

another person, in our contest the will-maker is the latter, that is so controlling that it undermines the will-maker's ability to act freely and voluntarily. A person making the charge of undue influence must prove it.

What exactly is a "beneficiary"?

A person who is the recipient of money or properties distributed by a will or trust. A beneficiary could be a husband, wife, son, daughter, grandchild, distant relative or institution; a dog can be a beneficiary and even a cat.

I wish to state in my will that if anyone contests it, they receive zero. Zip. Zilch. Nada. Is this possible?

You want an *in terrorem* clause.

What's an *in terrorem* clause?

In terrorem is Latin for "in terror." An *in terrorem* clause will state, generally: "If a person contests this Will, that person forfeits his or her inheritance." Some states permit this terrorizing. Some states prohibit this terrorizing.

I wish to leave my son my pocket watch? What do I do?

Make a bequest.

What is a "bequest"?

A gift under a will or a trust.

I wish to leave my swampland to my ex-wives, to share and share alike. What do I do?

Make a devise.

What is a "devise"?

A gift, usually of real estate, under a will or a trust. It can be used as a noun or a verb: "I devise my swampland to . . . " or "I received a devise under my boss's will, can you believe it?"

What is a "specific bequest" or a "specific devise"?

A gift of a particular amount or item of property. An example of a specific bequest: "I bequeath my false teeth to my son-in-law." An example of a specific devise: "I devise my toxic waste dump to my ex-partner."

What is a "general bequest" or a "general devise"?

A gift of unsegregated property such as "one-half of my stocks and bonds" or "one-half of my Florida real estate."

I wish to leave everything to my wife. How do I do this?

Give her your residuary estate.

What is an "estate" exactly?

An estate is the total of all assets (real estate, cash, bonds, stocks, automobiles, artwork, jewelry, partnership interests, etc.) that a person, alive or dead, owns.

What is a "residuary estate"?

The balance of a will-maker's property, after specific gifts and the payment of debts, taxes and expenses.

Can I change my will?

Sure. Sign a new will, revoking the old, or sign a codicil.

What is a "codicil"?

A written amendment to a will, executed with the same formalities of a full will (e.g., the signing in the presence of two witnesses and a notary public).

I have read "I give $10,000 to my issue *per stirpes*" in some wills. Who are "issue"? What is "*per stirpes*"?

Issue means children, grandchildren, great-grandchildren and more remote, direct lineal descendants.

Per stirpes means that if a child is predeceased, his or her children will receive his or her share of a bequest or devise in equal subshares.

What if I overlook someone — for instance, my spouse or one of my children — in making my will, but it's a stupid accident by my lawyer?

State laws address the situation of a pretermitted heir.

What is a "pretermitted heir"?

A survivor or descendant omitted by a will-maker

who, in many states, receives a share of an estate based on the distributions that would be made if the forgetful dead-head died without a will.

Why is a Last Will and Testament called a "Last Will and Testament"?

If you break the phrase down by words, it makes sense. "Last" is appropriate because usually your will consists of your last words. Immortality will be made easy by this wise-guy guidebook, as you will see. "Will" indicates your volition or desire. "Testament" means "creed" or "guidelines for the disposition of one's assets upon death." Together, "Last Will and Testament" is logical.

What is a "Testator"?

A man who makes and signs a will.

What is a "Testatrix"?

A woman who makes and signs a will. Some say a Testatrix has no testi . . . — never mind.

What is an "apportionment" clause?

It is the provision in a will that directs that death taxes be paid from certain assets or certain portions of inheritances. The apportionment clause can be important because it determines the net amount beneficiaries will receive after taxes.

I'm the proud owner. I'm leaving this to my ex-wife under my will.

CHAPTER THREE

TRUSTS MADE SIMPLE

What exactly is a "trust"?

A trust is an agreement, usually in writing and signed, where one person (or bank or trust company) holds property for another person according to the terms of the agreement.

What might a simple trust agreement say?

TRUST

I give $10,000 to my Trustee to hold for the benefit of my children as he deems appropriate.

____
Trust-maker's signature

____
Trustee's signature

Is the above trust valid?

The above would be a valid, binding trust; it's so crude it would be ambiguous but still legal. The form fails to address, among other things, who will serve as a successor if the named Trustee quits.

Why would anyone want a trust?

A trust may be desirable (a) when one does not want to give money in one big sum to a person, but rather in small installments, (b) to preserve money and protect beneficiaries, (c) for certain tax benefits, or (d) when one wants a person or company to hold assets for another based on a variety of other reasons to be discussed.

Is a trust a thing? What are its components?

A trust is a legal "thing" like a corporation or partnership. Its two basic parts, financially speaking, are principal and income.

What is a trust's "principal"?

The assets placed into a trust. The assets could be real estate, stocks, bonds, cash or anything else. People, even if they're involved in stocks and bondage, cannot be placed into trusts.

What is a trust's "income"?

The interest, dividends, rents or other earnings produced by the principal in a trust.

What is a "Trustee"?

A Trustee is the person or institution holding the property for another person or persons. A Trustee is responsible for distributing property in accordance with the trust's instructions. It can be a man, woman, bank or trust company; it cannot be a dog, even if the dog is your best, most trusted friend.

A Trustee should be trustworthy.

What is the "Grantor" or "Donor" or "Settlor" of a trust?

The individual who creates a trust.

Can I create a trust while I'm alive and kicking?

Yes. You're thinking of an *inter vivos* trust.

What is an *inter vivos* trust?

A trust created during your lifetime — in contrast to a trust provided within a will and established after you've gone underground. These answers, questions and answers are beginning to remind me of *Jeopardy.*

Inter vivos means "between living people." An *inter vivos* trust can be revocable or irrevocable, terms that will be explained in a matter of moments.

Can an *inter vivos* trust continue after I've entered the Pearly Gates?

Yes.

What is a "revocable trust"?

A trust that can be canceled or just changed by the trust-maker.

Why would someone choose a revocable trust?

Mostly because assets in a revocable trust generally need not be probated, upon death, and need not be involved in guardianship proceedings.

If a Grantor retains the power to revoke a trust, the revocable trust—the property held in the trust—continues to be controlled by the Grantor and owned for tax purposes by the Grantor. Principal may be subject to estate taxes upon the Grantor's death. Sometimes revocable or living trusts are called "Grantor trusts."

What is a "living trust"?

A trust, like Frankenstein, that takes on a life of its own. Actually, it's exactly the same as the revocable trust that I just defined.

If I set up a Grantor trust, can I be the Grantor, the Trustee and the beneficiary— all in one?

Yes, you lucky duck. A living trust is a legal fiction that is accepted widely. It is a popular means of avoiding probate (to be discussed in the suspenseful future). When you set up a grantor trust, you can use your own social security number and include its income on your personal income tax return so that it is no big deal, generally speaking, until beneficiaries other than yourself become involved. Alas, New Yorkers must name an additional Trustee other than themselves in creating Grantor trusts.

What is an "irrevocable trust"?

A trust that cannot be canceled. All trusts created in wills are irrevocable, unless the decedent returns from the dead for a rewrite.

Why would someone choose an irrevocable trust?

Mostly to remove an irrevocable trust's assets from his or her estate, reduce the size of an estate, lower death taxes and provide goodies to the beneficiaries of the irrevocable trust.

If a Grantor places property into an irrevocable trust and gives up control of the property and the trust, he makes a gift so that the property no longer belongs to him and, usually, is excludable from his estate.

I wish to set up a trust for my minor children. I would like income and principal to accumulate until they reach age twenty-one, at least. What kind of trust should I establish?

You may create a "2503(c) trust" or a "Crummey trust." Also, this is a good time to become familiar with "spendthrift" and "sprinkle" trusts.

What is a "2503(c) Trust"?

2503(c) is the section of the Internal Revenue Code in which this type of trust is addressed. It is a trust for a minor in which principal must be paid to the minor upon reaching age twenty-one. Gifts to the trust can be planned each year while the child is growing up. The I.R.S. makes certain guidelines. The income may be used for the minor during his minority; accumulated income and principal must be paid to the minor at age twenty-one; and accumulated income must be payable to his or her estate or by his will if the minor dies before reaching age twenty-one.

What is a "Crummey trust"?

A trust that is not worth a damn.

Actually, it's a trust usually used for children that can have flexible provisions for payments and, among other things, it need not terminate when the child reaches age twenty-one and is not necessarily able to handle money. Contributions can be made each year as the child is reaching maturity and a substantial sum can be accumulated. Because of a tax court case involving Mr. Crummey, however, contributions to the trust must be subject to a "withdrawal right" by the beneficiary. This right by the beneficiary is usually called a "Crummey power."

What is a "spendthrift trust"?

A trust that is purchased on sale.

Actually, it is an *inter vivos* or testamentary trust set up for someone who spends money foolishly. The spendthrift trust usually regulates the flow of funds to the improvident person in a gradual, planned manner and places the trust's assets beyond the reach of the beneficiary's ex-spouses, loan sharks, gambling buddies, parasites and creditors.

What is a "sprinkle trust"?

A trust that gives the trustee the discretion to distribute differing amounts of income or principal among various beneficiaries. The sprinkle trust is used sometimes when beneficiaries, for example children, have different needs to be met and the trust-maker, for example a father, does not mind giving them potentially unequal amounts of money.

Can I create a trust within my will?

Yes; you're thinking of a "testamentary trust."

What is a "testamentary trust"?

A trust created under the terms of a will so that it is established after the death of the will-maker.

How are trusts used to protect against creditors?

Joe Blow, Sr., sets up a trust for Joe Blow, Jr. Joe, Jr., as the beneficiary of the trust, has the right to receive only income. Joe, Jr.'s creditors generally cannot reach the principal in order to collect Joe, Jr.'s debts. Joe Blow, Jr., cannot blow it.

Can I protect my assets from creditors crawling all over me by putting them into a trust for my own benefit?

If you have set up the trust, funded it and did so deliberately to defeat debts, your efforts may be undone in court by your creditors. This question is in the area of "asset protection trusts" and is slightly different from the estate planning intended by this primer. Bankruptcy and creditor-debtor specialists can advise you best on these trusts. I hope you don't aspire to become a deadbeat. I want you to make and preserve millions.

I live in the Northeast. I am moving to the Southwest. Can I take my trust with me?

If it is revocable, yes. If it is irrevocable, it depends

upon the terms of the trust regarding *situs*. If the trust is silent about *situs*, it depends upon the law of the state you are leaving.

What is *situs*?

The "home" of a trust. While usually it is the trust's birthplace, trust documents and state laws can provide that a trust can move away from home.

I wish to set up a trust for my daughter. I want her to receive all income during her lifetime, but not touch principal. She can give principal away upon her death. What to do?

Give your daughter a "power of appointment."

What is a "power of appointment"?

The power to direct principal of a trust to various recipients, usually by specific provisions in a will. In this example, your daughter can write in her will: "I exercise the power of appointment in my daddy's trust for the benefit of my children and direct the Trustee to pay all principal and income to them."

Are there any other kinds of trusts?

Yes. For example, life insurance trusts, marital trusts and charitable remainder trusts. These are more complicated and will be discussed in the chapters covering tax-saving plans and charitable giving. That's all for now, folks, on the basics of trusts.

I stole the card from some woman, but her husband is letting me keep it because I spend less than she spends.

Simple Form of Living Trust

____

199___ REVOCABLE TRUST,

between

____

as Donor,

and

____

as Trustee

Dated: _______________________, 199___

______________________ 199_____ REVOCABLE TRUST AGREEMENT dated ________, 199____, between ________________ as Donor, and ________________ as Trustee.

1. **TRUST PROPERTY.** The Donor gives the property described in Schedule A (attached) to the Trustee as the principal of the Trust. The Trustee acknowledges the receipt of such property and agrees to manage and dispose of it, together with any additions and income, according to this Agreement.

2. **LIFETIME MANAGEMENT AND DISPOSITIONS.**

During the Donor's lifetime, the Trustee shall (a) invest the principal of the Trust as the Donor may direct from time to time by a written instrument signed by the Donor and delivered to the Trustee and (b) pay to or apply for the benefit of the Donor as much of the net income and principal of the Trust, both as the Donor shall direct by written instructions signed and delivered to the Trustee.

3. **DISPOSITION AT DEATH.** Upon the Donor's death, the principal and interest of the Trust shall be distributed according to the following terms.

A. If the Donor's spouse, __________________, shall survive the Donor, the principal of the Trust remaining at that time (after the dispositions in Subdivisions A and B of this Section) shall be distributed to such spouse outright.

B. If the Donor's spouse shall not survive the Donor, the principal of the Trust remaining at that time shall be distributed to the Donor's issue *per stirpes* as shall be living at that time.

4. **TRUSTEE PROVISIONS.**

A. Governing Law and Powers of Trustee. This Trust shall be governed by the laws of the State of ________________. The Trustee shall have all powers provided under the laws of the State of ________________________.

B. Successor Trustee. If for any reason ______________________ is unable to serve as Trustee, ____________________ shall serve as successor Trustee.

C. Waivers. Any law or local court rule now or hereafter in effect requiring qualification, administration or accounting by the Trustee to any court shall not apply to the Trustee and are hereby waived by the Donor.

5. **REVOCABILITY.** The Donor shall have the right to amend or revoke this Agreement at any time and from time to time by a written instrument, signed, acknowledged and delivered to the Trustee, provided that the Trustee's duties may not be increased nor compensation decreased without its consent.

6. **PAYMENT OF DEBTS, EXPENSES AND TAXES.** Following the death of the Donor and upon the request of the legal representatives of the Donor's estate, the Trustee shall pay to such legal representatives or directly to the appropriate person or entities from the principal of the Trust, such sum or sums as shall be certified to them by such representatives to be due for any part or all of (a) the debts of the Donor, (b) the expenses of the Donor's last illness and funeral expenses, (c) the expenses of administering the estate of the Donor, and (d) any estate, transfer or other inheritance taxes, including any interest or penalty thereon imposed by the United States or any state thereof or any foreign country or subdivision thereof as a result of the death of the Donor, without regard to the amount, if any, of any such taxes, including interest and penalties, which would otherwise be payable from the principal of the Trust under any applicable provision of law.

IN WITNESS WHEREOF, the Donor and Trustee has executed this Agreement as of the date shown above.

________________________ ________________________

Witness Donor

____

Witness

________________________ ________________________

Witness Trustee

____

Witness

STATE OF ______________________________

COUNTY OF ______________________________

We ____________________, ________________ and ____________________, the Donor, Trustee and the witnesses, respectively, whose names are signed to the attached instrument, were sworn, and declared to the undersigned officer that the Donor in the presence of the witnesses signed such instrument as the ____________________ 199___ Revocable Trust, that h___ signed, and that each of the witnesses, in the presence of the Donor and in the presence of each other, signed the Trust as a witness.

____

Donor and Trustee

____

Witness

____

Witness

Subscribed and sworn to before me by ______________ ________________________, the Donor and Trustee and by__________________ and________________, the witnesses, on ______________199___, all of whom personally appeared before me. ________________ is personally known to me or has produced satisfactory proof of identification. ______________________ is personally known to me or has produced satisfactory proof of identification. __________________ is personally known to me or has produced satisfactory proof of identification.

____

Signature of Notary Public

Commission Information and Stamp:

SCHEDULE A

(List of Assets; for example, "$50,000, 10-acre farm in Millbrook, New York, and 1,000 shares of Microsoft")

Simple Form of Will to Pour Over into Living Trust

Last Will and Testament
Of

Dated: ____________________, 199______

I,_______________________, a resident of_____________________, do hereby declare this as my Last Will and Testament and revoke all wills and codicils heretofore made by me.

ONE: A. I give all tangible personal property which I may own at the time of my death (including without limitation jewelry, art, household furnishings and effects, clothing and automobiles) to such person or persons as I may designate from time to time in a letter, addressed to my Executor.

B. To the extent that I shall not dispose of any such tangible personal property by a letter, I give the balance or all of my tangible personal property to my spouse, ________________________________, or if my spouse shall not survive me, to such of my children as shall survive me, to be divided among them as they shall agree. If my children shall be unable to agree upon such division, such property shall be divided among them as my Executor shall determine in his or her absolute discretion.

TWO: All the rest, residue and remainder of my estate, real, personal or mixed, of whatsoever nature and wheresoever situate, of which I shall die seized or possessed, or to which I shall be in anywise entitled at the time of my death, including all property over which I shall have a power of appointment or other power of disposal at the time of my death and including any legacies or devises which may lapse or be invalid or for any reason fail to take effect (herein referred to as my "residuary estate"), I give and devise to the Trustee of the ________________________________ 199___ Revocable Trust, to be disposed of in accordance with the terms of such trust.

I direct that my residuary estate be treated as an addition of property by me to such Trust and administered in accordance with the terms of such Trust, commingled with any other property held by the Trustee thereof. I do not intend to create a testamentary Trust of my residuary estate. Accordingly, I relieve the Trustee of the ______________________________ 199___ Revocable Trust from any duty to treat any property transferred to them in accordance with the provisions of this Article as subject to a testamentary Trust and they shall not be required to account therefore to any court.

THREE: A. I nominate and appoint my __________, __________________, as Executor. If ___________________ shall fail to qualify or, having qualified, shall cease for any reason to act as Executor, I nominate and appoint my __________________, ______________________________, as successor Executor.

B. I direct that no bond or other security shall be required of any Executor for the faithful performance of his, her or its duties, any law of any state or jurisdiction to the contrary notwithstanding. I further waive compliance with any other law or local court rule now or hereafter in effect requiring qualification, administration or accounting by my Executor to any court.

FOUR: My Executor shall have all powers provided under the laws of the State of ___________________________.

FIVE: Under the provisions of the _______________ _______ 199___ Revocable Trust, the Trustee therein is directed, upon the request of the legal representatives of my estate, to pay such amounts as shall be needed to satisfy (i) my debts, (ii) the expenses of my last illness and funeral expenses, (iii) the expenses of administering my estate and (iv) any estate, transfer or other inheritance taxes imposed by the United States or any state thereof or any foreign country or subdivision thereof by reason of my death. It is my intention that the principal of the ______________________199___ Revocable Trust shall be used for the payment of such debts, expenses and taxes and I hereby direct my Executor to require the Trustee under the _______________________________ 199___ Revocable Trust, in accordance with the provisions thereof, to pay such amounts as shall be needed to satisfy any such debts, expenses and taxes.

IN WITNESS WHEREOF, I, ______________________, have set my hand this ____ day of __________________, One thousand nine hundred and ninety-________.

______________________ ______________________

Witness [Will-maker]

____

Witness

STATE OF ______________________________

COUNTY OF ___________________________

We, ______________________, and______________,
the Testat___ and the witnesses, respectively, whose names are signed to the attached instrument, were sworn, and declared to the undersigned officer that the Testat____ in the presence of the witnesses signed such instrument as h___ Last Will, that h___ signed, and that each of the witnesses, in the presence of the Testat____ and in the presence of each other, signed the Will as a witness.

____
Testat___

____
Witness

____
Witness

Subscribed and sworn to before me by____________ the Testat___ and by _________ and_______________ the witnesses, on ________________, 199__ , all of whom personally appeared before me. ________________ is personally known to me or has produced satisfactory proof of identification. _________________________ is personally known to me or has produced satisfactory proof of identification. ______________________ is personally known to me or has produced satisfactory proof of identification.

____
Signature of Notary Public
Commission Information and Stamp:

Simple Form of Spendthrift Trust for Children

______________________________199__ TRUST

between

____

as Donor,

and

____

as Trustee

Dated: ______________________, 199__

TRUST AGREEMENT dated ____________, 199____ between ____________ as Donor, and ________________ as Trustee, for the benefit of the Donor's child, ____________________, (the "Beneficiary").

1. **TRUST PROPERTY.** The Donor gives the property described in Schedule A (attached) to the Trustee as the principal of the Trust. The Trustee acknowledges the receipt of such property and agrees to manage and dispose of it, together with any additions and income, according to this Agreement.

2. **DISPOSITION OF INCOME AND PRINCIPAL.**

A. The net income of the Trust shall be paid or applied for the benefit of the Beneficiary in convenient installments, but not less often than annually, as the Trustee shall determine in its absolute discretion, provided, however, that the Trustee may accumulate and add to principal all or any part of such net income, any such capitalized income thereafter to be disposed of as a part of such principal.

B. The Trustee is authorized, at any time and from time to time, to pay or apply any part or all of the principal of the trust to or for the benefit of the Beneficiary for any reason whatsoever, even though any such payment or payments may result in the termination of the Trust.

SELECT AN ALTERNATIVE CHOICE VI, VII or VIII:

CHOICE VI:

OUTRIGHT DISTRIBUTION OF PRINCIPAL TO CHILD AT AGE ______:

C. When the Beneficiary shall attain age ________, the principal of the Trust at that time remaining shall be distributed to the Beneficiary.

D. Upon the death of the Beneficiary before attaining age _____, the principal of the Trust, at that time remaining, shall be distributed to such of the issue *per stirpes* of the Beneficiary as shall be living at that time or, if none, to such of the issue *per stirpes* of the Donor as shall be living at that time, provided, however, that any property distributable to a child or more remote issue of the Donor for whom a Trust created by the Donor shall be in existence at that time shall be added to the principal of such Trust, to be disposed of as a part thereof.

CHOICE VII:

DISTRIBUTION OF PRINCIPAL TO CHILD IN 3 STAGES:

C. When the Beneficiary shall attain age __________, one-third (1/3) of the principal of the Trust at that time remaining shall be distributed to the Beneficiary.

D. When the Beneficiary shall attain age __________, one-half (1/2) of the principal of the Trust at that time remaining shall be distributed to the Beneficiary.

E. When the Beneficiary shall attain age __________, the principal of the Trust at that time remaining shall be distributed to the Beneficiary.

F. Upon the death of the Beneficiary before attaining age __________, the principal of the Trust at that time remaining shall be distributed to such of the issue *per stirpes* of the Beneficiary as shall be living at that time or, if none, to such of the issue *per stirpes* of the Donor as shall be living at that time, provided, however, that any property distributable to a child or more remote issue of the Donor for whom a trust created by the Donor shall be in existence at that time shall be added to the principal of such trust, to be disposed of as a part thereof.

CHOICE VIII:

LIFETIME TRUST FOR CHILD:

C. Upon the death of the Beneficiary, the principal of the Trust at that time remaining shall be distributed to such of the issue *per stirpes* of the Beneficiary as shall be living at that time or, if none, to such of the issue *per stirpes* of the Donor as shall be living at that time, provided, however, that any property distributable to a child or more remote issue of the Donor for whom a trust created by the Donor shall be in existence at that time shall be added to the principal of such Trust, to be disposed of as a part thereof.

Optional Crummey Power if gifts are contemplated during Donor's or Donors' lifetime:

___________. If at any time any addition by way of gift is made to the trust, the Beneficiary (or, if he or she shall be a minor, his or her Guardian) shall have the right, at the time or times and in the manner set forth in this Subdivision, to withdraw from the trust an amount equal to such addition but in no event to exceed such amount as shall be equal to the annual gift tax exclusion available to the transferor of such addition in the year of such addition pursuant to Section 2503(b) and 2513 of the Internal Revenue Code, as amended from time to time (the "Code"). The Trustee shall give reasonable notice of any such addition and the amount thereof to the person who, at the time of such addition, is entitled to withdraw such addition. Such right of withdrawal shall be exercisable only by written demand addressed to the Trustee by the person exercising such right of withdrawal within thirty (30) days after such person receives such notice from the Trustee, and the amount so demanded shall be paid by the Trustee promptly upon receipt of such notice. The right to demand payment pursuant to this Subdivision shall not be cumulative and to the extent that such right of withdrawal is not exercised by the end of such thirty (30) day period such right of withdrawal shall lapse. The Trustee shall retain any such addition to this Trust until the end of such thirty (30) day period, in the form received or as converted into cash or marketable securities.

3. TRUSTEE PROVISIONS.

A. Governing Law and Powers of Trustee. This Trust shall be governed by the laws of the State of ______________.

The Trustee shall have all powers provided under the laws of the State of _____________.

B. Successor Trustee. If for any reason __________ __________ is unable to serve as Trustee, _______________ shall serve as successor Trustee.

C. Waivers. Any law or local court rule now or hereafter in effect requiring qualification, administration or accounting by the Trustee to any court shall not apply to the Trustee and are hereby waived by the Donor.

4. **ADDITIONAL PROPERTY.** If at any time additional property of any kind acceptable to the Trustee shall be assigned and transferred to them by the Donor or any other person, *inter vivos* or by will admitted to probate, or delivered to the Trustee to be added to the Trust, the Trustee shall accept such property and hold the same as part of the Trust estate, subject to the terms and conditions of this Agreement.

IN WITNESS WHEREOF, the Donor and Trustee have executed this Agreement as of the date shown above.

_______________________ _______________________
Witness , Donor

____
Witness

_______________________ _______________________
Witness , Trustee

____
Witness

STATE OF ______________________________
COUNTY OF ____________________________

We,___________________, ____________________ and_____________________________, the Donor, Trustee and the witnesses, respectively, whose names are signed to the attached instrument, were sworn, and declared to the undersigned officer that the Donor in the presence of the witnesses signed such instrument as the ________________ 199___ Revocable Trust, that h___ signed, and that each of the witnesses, in the presence of the Donor and in the presence of each other, signed the Trust as a witness.

____
Donor and Trustee

____
Witness

____
Witness

Subscribed and sworn to before me this _____ day of __________,199___ by ________________, as Donor, and _________________, as Trustee. ___________________ is personally known to me or has produced satisfactory proof of identification. ___________________ is personally known to me or has produced satisfactory proof of identification.

____
Signature of Notary Public
Commission Information and Stamp:

SCHEDULE A

(List of Assets; for example, "$100,000")

CHAPTER FOUR

FIDUCIARIES: EXECUTORS, TRUSTEES AND LAWYERS

What is a "Fiduciary"?

A Fiduciary is a person or organization, such as a bank or trust company, given the responsibility to act for the benefit of another.

Is an Executor a Fiduciary?

Yes, an Executor is a Fiduciary named in a will who acts for the benefit of the deceased, beneficiaries named in a will and others, such as creditors and tax collectors.

Is a Trustee a Fiduciary?

Yes, a Trustee is a Fiduciary named in a trust agreement who is supposed to serve the interests of the beneficiaries of the trust.

Is a Guardian a Fiduciary?

Yes, a Guardian is a Fiduciary who is duty-bound to protect a child or an adult who is incapacitated by a disability or age.

Is an attorney a Fiduciary?

Yes, an attorney is a Fiduciary with duties to his or her client. The client may be a will-maker or trust-maker or also a Fiduciary. Executors, Trustees and Guardians usually are represented by lawyers.

Are Fiduciaries important?

Yes, extremely important. One can create the most intelligent will or trust, but a wrong-minded Fiduciary may fail to carry out its terms. Fiduciaries are in a superior position to lie, steal and cheat if they are so inclined.

What happens if a Fiduciary lies, steals or cheats?

Fiduciaries have duties to their beneficiaries. If they fail to discharge their duties, they can be sued. If a Trustee makes idiotic decisions in investing trust assets, the Trustee can be forced to reimburse the trust for its losses. If an Executor buys an estate's asset at one-half its actual value, the Executor can be removed in probate proceedings and forced to return the asset to the estate. Attorneys who breach fiduciary duties to their clients can be sued and reported to their state bar association for disciplinary action.

How should I best select a Fiduciary?

The selection of a Fiduciary requires careful thinking. It is nice when one appoints his or her spouse to act as a Fiduciary and this is the predominant pattern in our society. Adult sons, daughters, grandsons and granddaughters can be good choices. Banks and trust companies are professional Fiduciaries that are popular selections, although many people regard them as impersonal and expensive. Banks and trust companies, sometimes referred to as institutional Fiduciaries, usually have the advantage of deep pockets that enable them to pay for inevitable mistakes.

In selecting a Fiduciary, I recommend that you have a frank discussion with the prospect as if you are interviewing a person for employment. In fact, a Fiduciary is an employment choice. Ask the prospect a series of questions to satisfy yourself that your wishes will be followed and your ideas properly understood. The trustworthiness of the prospect is imperative. Even if the prospect is not extra- ordinarily intelligent or sophisticated, he, she or it may be honorable and able to hire professionals (e.g., investment advisors, accountants, etc.) to help in the discharge of fiduciary duties.

No one is perfect, of course, and no Fiduciary will be.

Are Fiduciaries paid?

Almost always. Some may waive their fees; wave them good-bye. While family members sometimes forsake fees, others rarely do. Since Fiduciaries often have a lot of work to do, they deserve to be compensated.

How are Fiduciaries paid?

Most will accept money.

All fiduciary fees are negotiable. In some states, commissions payable to Executors, Trustees or lawyers are fixed by statute. Some statutes establish fees according to a schedule; for example, an attorney may be paid 3 percent of the value of probate assets. Other statutes require that fiduciaries be paid "reasonable compensation" based on certain criteria, such as the amount of time involved, the complexity of the work, the risks involved and the special skills of the individual or institution.

Can I appoint more than one individual to serve as Fiduciaries? Can I appoint an individual and an institution?

Yes. Yes. These may insure that your wishes are carried out in that the "cofiduciaries" may bring complementary talents to their duties and check each other. On the other hand, the dark view, the coexecutors or cotrustees can conspire to lie, cheat and steal together and their misdeeds made more difficult to discover. Just don't die and everything will be all right.

What is a Fiduciary able or authorized to do?

Fiduciaries are given specific powers within wills and trusts. In addition, state laws spell out what a Fiduciary may or may not do. Some actions by Fiduciaries require court approval. In Florida, an Executor cannot sell real estate without court approval unless specifically authorized under a will or trust document.

If a Fiduciary makes a mistake, can he be made to pay for the mistake out of his own funds?

Yes. Fiduciaries bear personal responsibility for their conduct. Usually they are O.K. if they follow the "prudent investor rule."

What is the "prudent investor rule"?

A Fiduciary is obligated to use such care and skill as an ordinary, prudent investor would exercise with respect to his own property.

The prudent investor rule formerly was the "prudent man rule." This prevented a woman from investing prudently, so the term was changed.

Can a Fiduciary buy assets of an estate for himself?

He had better do this with extreme care; otherwise, he can be accused of improper "self-dealing."

What is "self-dealing"?

As the term suggests, actions by a Fiduciary that benefit himself. If an Executor, for example, buys an automobile from an estate at one-half price — that's improper self- dealing. If he buys it at full price with the consent of all interested parties — that's usually considered kosher.

How can I find an ethical lawyer?

In order to find the best lawyer possible, interview a lot of them, and after all of several interviews are finished, pick the one that seems most honest. Honesty, intelligence, expertise and empathy are probably the best criteria for evaluating legal parasites. The order of these three measures depends on your needs.

What does a Trust Officer really do?

According to one comedian, he manages money until it's all gone.

My parents left my inheritance in a spendthrift trust. They didn't expect you to be my Trustee.

CHECKLIST 4:

INTERVIEW WITH THE VAMPIRE: QUESTIONS TO ASK LAWYERS BEFORE HIRING THEM

1. How many years have you been practicing law?
2. How many wills have you written?
3. Have you ever been sued for malpractice? If no, you're lucky. If yes, please describe the circumstances (in other words, why it was everyone else's fault).
4. What do you charge?
5. Is it difficult to reach you by phone?
6. Do you delegate your work to other attorneys or paralegals? If yes, tell me about them and how you charge for their time.
7. What distinguishes you, if anything, from other lawyers?
8. What do you propose to do if you and I have a dispute?
9. Where did you attend college? Law school? Where did you learn to write wills?
10. Describe your experience in audits by the I.R.S. and in probate litigation.
11. Has any will or trust you have written been contested? If yes, please describe the situation.
12. Do you mind if I get second legal opinions to check yours? If yes, what are you afraid of?

CHAPTER FIVE

PROBATE AND PROBATE PROBLEMS

What is "probate"?

Probate, generally speaking, is the process in court by which a judge determines the validity of a will and makes certain that the terms of the will are carried out, that debtors and tax collectors are paid and disputes are resolved among family members or beneficiaries.

Isn't probate just plain bad?

No. Many people misperceive probate as bad because in probate court nasty battles have been waged among current and ex-wives, stepchildren, nurses, leeches and others who may act like vultures over the corpse's corpus. In actuality, most probate proceedings run very smoothly and simply. Probate is not probation.

Are there any advantages to probate?

Yes. First, it is a way of determining, once and for all, the creditors of the deceased and either making final payment to them or forever stopping their claim. Second, it involves the oversight of a judge who, at the nominal cost of filing fees, makes certain that the Executor acts honestly and that the will is followed properly. The judge may make sure that Executors and lawyers do not overcharge or commit misdeeds, such as selling probate assets to themselves at bargain-basement prices. Third, probate in and of itself does not cause any death taxes and involves relatively little expense (unless someone, such as the Executor or a shyster, overcharges).

Are there any disadvantages to probate?

Yes. First, the disposition of one's estate usually becomes a matter of public record — so there is a loss of privacy. Second, probate often is more time-consuming than the "automatic" administration of living or revocable trusts. Third, probate may involve some attorneys' fees, which although often exaggerated, can be saved. Fourth, probate can involve a lot of paperwork. Fifth, probate procedures sometimes can delay the distribution of legacies.

When Norman F. Dacey, a nonlawyer, published a how-to book entitled *How to Avoid Probate!* many people began to believe that probate is awful and ought to be avoided. Judge the advantages and disadvantages for yourself.

What happens in probate court if I die without a will?

A judge picks an Executor or Executrix. Usually, spouses and family members receive preference, but the judge may pick a stranger, a bank or even a strange bank.

The dead person's dough is distributed according to the laws of the state in which his or her permanent residence was located. Half of the assets may go to the surviving wife and the other half in equal shares to his kids. Creditors and taxes must be paid.

Presidents Lincoln, Johnson, Grant and Garfield died without wills. Howard Hughes died without a will. Pablo Picasso also died without one; one of his paintings was entitled "Will," but no one could interpret it, and his signature was not witnessed or notarized.

What is "intestacy"?

It has nothing to do with ecstacy. It has nothing to do with testosterone. It is the situation just discussed, where a person dies without a will and his estate is managed by a court-appointed Executor and divvied up by state law.

What happens if I deliberately omit my husband from my will?

In most states, an omitted spouse has a right to some inheritance called an "elective share."

What is an "elective share"?

The right a husband or wife has to his or her deceased spouse's money. It can be 30 percent, 50 percent or some other percentage.

If my husband must receive, say, 30 percent of my estate, can I leave it to him in a trust that pays income to him during his lifetime and upon his death passes to my children?

In some states, this arrangement is an allowable provision for the elective share. In other states, your husband would be entitled to his elective share outright and not in trust. You may wish to change states; it sounds like you may also wish to change husbands.

How long does probate last?

The shortest probate proceeding that I know of lasted nine months; the decedent could have had a baby. The longest that I know of lasted forty-four years. The time depends on whether there are disputes, the efficiency of lawyers and judges and Executors, the complexity of the will and many other factors.

Is the entire estate of the deceased reported to the probate court?

Assets in the deceased's name alone, usually, are those which must be reported in an inventory submitted to the probate court. These assets "pass under the will." In contrast, another asset may be owned jointly with a person who survives, and the survivor takes the entire asset automatically by operation of law.

What are the basic steps in probate?

First, the will is submitted to a judge, and if it is determined to be valid, it is "admitted to probate." Second, an executor is appointed. Third, notice of probate administration is published in a local newspaper and, usually, mailed to known creditors of the deceased. Fourth, an inventory of assets held in the deceased's name is filed in court. Fifth, claims of creditors are examined — if the claims are valid, they're paid; if the claims are invalid, they're not paid. Sixth, beneficiaries begin to receive part or all of their inheritances. Seventh, after federal and state taxing authorities confirm that all death taxes have been paid, a final accounting of monies received and distributed is filed. Eighth, if all has gone well, the probate judge discharges the Executor or Executrix from further responsibilities and the estate is closed. Obviously, will contests can change everything.

Probate seems complicated. Can I avoid it?

Yes, by two basic methods: revocable or living trusts are one way; joint ownership arrangements are a second way.

What are "revocable" or "living" trusts?

You can set up a revocable or living trust, put all of your assets into the trust, and the assets will be transferred automatically upon your death in accordance with the terms of the trust. I discussed these in Chapter Three. Some people think living trusts are panaceas in avoiding probate. These revocable trusts also can enable an incapacitated person to avoid guardianship proceedings — a cumbersome court process where a guardian is appointed by a judge to take care of a person and/or a person's property and make annual reports. Guardianship may be unnecessary where a living trust provides that a certain Trustee takes care of everything in the event the trust-maker becomes crazy or comatose or otherwise incompetent.

What is "joint ownership"?

You can put all of your assets into joint names, with rights of survivorship, using your name and the name of the person to whom you wish to give the assets. Upon your death, assets will transfer automatically to the surviving joint owner. Please understand the terms "joint ownership with rights of survivorship," "tenants by the entirety" and "tenants in common."

What is "joint ownership with rights of survivorship"?

The co-ownership of property, of any kind, by two or more people in which title of ownership (for example, the deed to real estate) indicates that upon the death of one co-owner, the surviving co-owner becomes the sole owner of the property.

If an asset is owned jointly with rights of survivorship, it is a winner-take-all proposition. Stay fit and survive so you will win the asset.

Joint ownership, by the way, has nothing to do with the possession of marijuana.

What are "tenants by the entirety"?

Co-ownership by a husband and wife of an asset with rights of survivorship. It's strictly spousal.

What are "tenants in common"?

Co-ownership of property without rights of survivorship. When one co-owner dies, his or her interest becomes part of his or her estate rather than transferring automatically to the surviving co-owner.

When, where and how do I address these different forms of co-ownership?

In the title to your house, cars, savings accounts, checking accounts and other assets, you can designate the form of co-ownership.

For example, bank tellers will have you check a box opposite "joint ownership with rights of survivorship" or "tenants in common." When you and your spouse buy a home, the warranty deed can state "John Doe and Jane Doe, as Tenants by the Entirety."

What happens if I set up a revocable trust or the joint ownership of assets with rights of survivorship, but I forget about an asset and it remains in my name alone at the time of my death?

The particular asset must be probated. When using

a revocable trust, it usually is a good idea to sign a "pour-over will."

What is a "pour-over will"?

A simple will by which you leave all of your assets to your revocable trust. If you transfer all of your assets into your revocable trust or joint ownership with survivorship rights, the pour-over will not need to be probated. It is just a precautionary measure. Many people with the best of intentions of avoiding probate fail to do so because they overlook the transfer of some asset, such as a savings bond hidden in a book or uncashed checks stashed in a cookie jar.

What does probate have to do with estate taxes?

Almost nothing. Some people are under the mistaken impression that one avoids death taxes if one avoids probate. This is false; a vicious lie. The I.R.S. will seek estate taxes, if owed, whether the deceased left a will, revocable trust, jointly owned checking accounts or just gold coins. Never underestimate the I.R.S.; they can get money from a corpse.

My husband is leaving me everything under his will. If I kill my husband, will I still receive my inheritance?

In most states, convicted murderers forfeit the inheritances they would otherwise receive from their victims. Sorry about that, sweetness.

We own this joint, but with no rights of survivorship.

CHECKLIST 5

THE PROBATE PROCESSIONAL

This schedule of events gives you some sense of probate. Probate differs from state to state, but this should give an overall picture.

PROBATE ADMINISTRATION SCHEDULE

Estate of [Dead Person]______ Date of Death: __________

	Deadline	Date Done
1. File Petition for Probate Administration, attaching original Will and Death Certificate, with Probate Clerk.	________	________
2. File Oath of Executor (by which he swears to do his duty), Designation of Resident Agent (by which a person is named to accept the delivery of documents if a claim or lawsuit is filed against the estate) and Acceptance (by Resident Agent) with Probate Clerk.	________	________
3. File proposed Order Admitting Will to Probate and Appointing Executor. Judge will sign Order (if this is in order).	________	________

	Deadline	Date Done
4. File proposed Letters of Administration (the "license" for the Executor to act in place of the dead person). Judge will sign (if all is in order).	________	________
5. Notice of Probate Administration:		
A) Publish Notice once a week for 2 consecutive weeks in a local newspaper to notify the public, in general, and potential creditors, that the dead person is dead; long live the Executor.	________	________
B) File Proof of Publication of Notice (from the newspaper) with Probate Clerk.	________	________
C) Send Notice of Probate Administration to surviving spouse, all known beneficiaries and all known creditors (CertifiedMail/ Return Receipt Requested).	________	________
D) File Proof of Service of Notice of Probate Administration (with return receipts from spouse and beneficiaries).	________	________

	Deadline	Date Done
6. Obtain appraisals of all real estate and personal property that the corpse owned.	______	______
7. Send Letter: A) Banks: Obtain statements from banks stating date of death balances for checking, savings and money market accounts.	______	______
B) Obtain valuation of any U.S. Savings Bonds; redeem.	______	______
C) Stockbrokers: Cancel any open orders to buy or sell stocks, bonds or options; inquire about accounts; request value of stocks and bonds as of date of death; advise brokers and companies of any address change.	______	______
D) Life Insurance: Send policy to life insurance company for payment and obtain Form 712 for I.R.S.	______	______
8. File Inventory (by which all probate assets are reported to the court) within 60 days after Letters of Administration (or obtain an extension).	______	______

	Deadline	Date Done
A) List all assets with estimated value; Executor must sign.	_______	_______
B) Send Inventory to surviving spouse and all known beneficiaries	_______	_______
C) File supplemental Inventory immediately after discovery of additional assets, if any, and serve copies as required for regular inventory.	_______	_______
D) File for extension if needed.	_______	_______
E) File Proof of Service of Inventory.	_______	_______
9. Obtain federal tax I.D. number for estate.	_______	_______
10. Confirm preparation of income tax returns by accountant.	_______	_______
11. Claims against estate: pay and file satisfactions of claim. Alternatively, don't pay and file objection to claims.	_______	_______
12. Objection to claim: An objection must be filed on a timely basis or the ability to object may be lost.	_______	_______

		Deadline	Date Done
13.	Pay state and local taxes of any kind as they become due. The estate is a taxpaying entity.	______	______
14.	Revalue assets as of 6 months after date of death.	______	______
15.	Obtain tax I.D. number for trusts, if any.	______	______
16.	File Notice of Federal Tax Return form within 12 months from the date letters were issued.	______	______
	Copies of the notice shall be served on interested persons. Whenever the due date is extended subsequently, similar notice must be filed and served.	______	______
17.	File with the I.R.S. and state of domicile, Federal Estate Tax Return Form 706, with certified copies of will and death certificate and other estate tax returns as needed. Send any payments owed. Due date: 9 months from date of death.	______	______
18.	File Proof of Payment of Estate Taxes (federal, state, other form).	______	______

	Deadline	Date Done
19. Complete distributions to Beneficiaries.	______	______
20. File Final Accounting form (serving copies to interested parties) or Waivers of Final Accounting form.	______	______
21. File Petition for Discharge of Executor form (from duties and liabilities) to close out the probate proceeding.	______	______
22. Obtain Order of Discharge form from probate judge.	______	______

CHAPTER SIX

COMMON PROBLEMS AND PRACTICAL SOLUTIONS

I am seventy years old and I'm dating a thirty-five-year-old woman. How can I get her to marry me?

Tell her you're ninety.

What the hell is wrong with paying estate taxes?

Absolutely nothing; it's an American right. Indeed, Supreme Court Justice Oliver Wendell Holmes wrote that "taxes are the way we pay for a civilized society."

If this is your viewpoint, you may disregard all of the spiffy tax-saving advice in this book.

I am a divorced man with $1,000,000 and two children. I have a girlfriend whom I wish to marry, but I wish to pass most of my money to my children when I die. What are my options, O great suggestor?

The first option is for you to enter into an antenuptial agreement, also known as a premarital contract, with your wife-to-be. In this agreement, you and she can establish almost all of your monetary obligations to one another. You can even etch in stone the sexual relationship you will maintain.

The second option is not to marry but just live together or go on living together. A cohabitation agreement is suggested, but not a necessity. A cohabitator does not have inheritance rights like a spouse.

The third option is to marry, include your new wife in your will and leave her the amount equal to her inheritance right under the laws of the state in which you both live. In Florida, for instance, this would be 30 percent; the balance could go to your kids.

The fourth option is to marry and create a trust for your spouse by which she receives income for life, and then everything springs to your offspring after she dies.

I have two sons, Cain and Abel. Guess what? Cain is bad and Abel is good. Cain is an alcoholic, cocaine addict, womanizer, goof-off and has every other vice available in America. Abel is a hardworking family man who attends church every Sunday and lives a conservative lifestyle. What should I leave them?

In the final analysis, whatever makes you feel good and fair.

You could leave them equal amounts of money, but give Abel his inheritance outright and place Cain's inheritance in a spendthrift trust with incentives for good conduct.

Warren Buffett, the wholesome Midwestern billionaire, is opposed to providing children with a "lifetime supply of food stamps just because they came out of the right womb." As a different viewpoint, Joseph P. Kennedy supposedly gave each of his children $1,000,000 with no strings attached when they reached age twenty-one so that they could devote their lives to public service.

Choices in wills are personal and there are no hard-and-fast, right-or-wrong answers.

I do not want immortality. I don't even want to be hooked up to one of those life-support systems that keep people alive after they have become veggie burgers. What do I do to prevent this?

Sign a "living will."

What is a "living will"?

This is a document, different from state to state, by which you declare that you do not wish to be kept alive by artificial means after you have reached a terminal condition.

Make sure your doctors determine that your condition is terminal, not your enemies. You're going to have big problems if your doctors are your enemies.

Read the book *Everything You Always Wanted to Know about Medicine, Sex and Death, But Didn't Want to Pay a Doctor to Find Out.*

A form of a living will follows the questions and answers of this chapter.

I am less concerned about death than old age. What if I lose my mind? What can I do to protect myself?

One advantage to living trusts is that you designate a successor Trustee to manage all of your trust's assets if you become mindless. A successor Trustee should be a person who looks after not only your assets but your health and welfare. Choose carefully.

If you don't do anything, your family or friends may be forced to seek a court-selected Guardian to manage you and your money. This can be very expensive, public and unpleasant. So do something.

If you have a spouse or child or friend you trust, you may give this individual a "durable power of attorney."

What is a "durable power of attorney"?

A durable power of attorney enables a designated individual to act on your behalf if you become incapacitated. It is easy to give durable powers of attorney.

A durable power of attorney form is located at the end of this chapter; check it out.

I have earned all of the money in my family. Neither my wife nor my children are knowledgeable about managing money. Do you have any special suggestions about how it can be handled after my death?

Your selection of a Fiduciary will be especially important, since a Fiduciary — a trustworthy, sophisticated individual, bank or trust company — seems needed to make the decisions you presently make. You pick stocks, manage real estate, select certificates of deposit, etc.; they can do these things.

In addition to carefully selecting the Fiduciary, you should be the one who negotiates a fee agreement, in writing, with the Fiduciary before you become incapacitated or die. You obviously can handle the negotiations better than the family members you describe.

I wish to leave different amounts to my children because they have different needs. One is rich, one is middle class, one is poor. What suggestions?

You many not have created all of your children equally.

Many people leave their children different amounts of money. Nothing is wrong with this. However, you may wish to make an ever-lasting expression of your equal love for your kids through an "ethical will."

What is an ethical will?

An ethical will is a letter to your loved ones, to be opened after your death, by which you may confirm your love for them, express sentimental thoughts, give advice or state anything else that you want them to keep in their heads and hearts.

I have children and grandchildren that have different needs. How can I meet their different needs?

The needs of offspring are almost always unpredictable. It seems that a "sprinkle" trust, with a good Fiduciary, will enable different amounts of money to be given to your children and grandchildren as circumstances change over time. One child may need money to buy her first home; another may need money after a divorce.

Three separate, equally funded trusts, but with different payout provisions for three different kids, may be considered instead of one sprinkle trust.

Since my second wife and my children by a former marriage do not get along, I anticipate a will contest. Is there anything extra I can do to prevent this?

As you know, anyone can sue anyone else over anything.

If you sign a valid will and give proper consideration to elective share laws, which entitle surviving spouses to a certain automatic share of an estate, you should avoid a battle.

To diminish claims of your incapacity at the time you signed your will, as well as claims about coercion, duress and undue influence, you may wish to have your will-signing recorded on videotape and give a statement in front of the camera to explain your voluntary desires and prove that you have all of your mental faculties.

When signing your will, you could have two doctors act as your witnesses. They later can serve as witnesses to your "sound mind." By the way, did you know that you can be of sound mind even if you have lost your hearing?

I am a sixty-year-old widow. I have one daughter and two grandchildren. My daughter and I do not get along. I do have an excellent relationship with my grandkids and want to provide for their college educations. What do you recommend?

It seems that you and your grandchildren may have a common enemy — the generation in between you.

If you are secure financially and have excess money, you should consider a lifetime, irrevocable trust for the benefit of your grandchildren and put up to $10,000 (increasing after 1998 for the cost of living) each year into it for each grandchild. You may wish to use a part or all of your unified credit (the $625,000, increasing gradually to $1,000,000 by 2006, that you can give free of gift taxes) if you can afford to part with this money.

For the really rich, however, please note that amounts given in excess of $1,000,000 are subject to generation-skipping taxes. Everyone has a $1,000,000 generation-skipping tax exemption.

If you are not financially secure and have no excess income, you may wish to set up within your will a trust — a testamentary trust — that provides for your grandchildren, including their education, medical and other needs.

We are a young couple with minor children. If anything happens to us, we are concerned about their well-being. What special measures should we take?

Life insurance may be a good place to start, unless you are very rich. You may consider setting up an irrevocable trust, put up to $20,000 per year per child into the trust, and have the Trustee purchase life insurance on your lives.

In your wills, I encourage you to name Guardians for your kids. This crucial choice requires a lot of thought over who will take care of them in the way that you want. You may wish to write a letter to your selected Guardian, or Guardians, with your special aspirations for your kids — a religious upbringing, a desire that they obtain a college education, a hope that they become exactly like you. Or unlike you, if you have doubts about yourself.

I have $800,000, from an inheritance, and my husband and I together have $50,000 in a joint savings account. We have three children. How can we leave them the most money in our wills?

You may wish to make a gift of $200,000 to your husband and change the joint account into an account in his name alone. Neither of these transfers are taxable under the unlimited marital deduction discussed elsewhere. You will then have $625,000 and he will have $250,000. This change obviously must take place after considering the stability of your marriage and your spouse's trustworthiness.

If you each set up a credit-shelter trust or by-pass trust,

all $850,000 ultimately will pass to your children and not one cent of estate taxes will be owed. These trusts are explained in the next chapter.

? I would like to set up a trust for my children, grandchildren, great-grandchildren and great-great-grandchildren so that the assets I have accumulated remain intact and they all receive just income. Is this possible?

Yes, but like almost everything, there are some restrictions and complications.

Generation-skipping taxes, for amounts over $1,000,000, should be considered.

Many states have a "rule against perpetuities" that prevents trusts from continuing forever. Trusts usually must terminate twenty-one years after the end of the lifetime of someone who is living at the time of your death. The State of Delaware allows a trust to continue forever and you may try to use this state as the *situs*, or home, of the trust.

Some people call these types of trusts "Dynasty Trusts," not after the television series, but rather because they often involve large sums of money and families like the DuPonts, Rockefellers, Phipps and Kennedys.

? I own a small business and wish to leave it to my sons. The estate taxes on the company alone at my death could force them to sell it. Any suggestions?

Of course; if I am anything, I am a suggestor.

You should begin a gift-giving program to your sons

of fractional interests in the company. This not only will reduce the size of your estate, but reduce the value of any interest retained by you and included in your estate. This is because fractional interests are allowed significant discounts recognized by the I.R.S.

Obviously, you may wish to control the company until the day you die; this is a separate, psychological subject.

Sophisticated attorneys can tell you all about family asset freezes. See the next chapter for a brief discussion.

If I move to the Bahamas with all of my money, will I avoid estate taxes?

Expatriation is an estate-planning option, even if it is unpatriotic.

Practically speaking, the I.R.S. may have trouble finding your corpse and your money. Legally speaking, regulations addressing expatriation make it very difficult. Expatriation ain't such a good idea.

I am married to a foreigner. Does this create any special problems?

Yes. If you die first, your spouse does not receive the unlimited marital deduction unless you set up a qualified domestic trust.

What is a qualified domestic trust?

A trust for your spouse for life, with a U.S. citizen or domestic corporation as Trustee, and other special requirements to insure that taxes are eventually paid to the U.S.

I don't think we should bury the jewelry.
Quick, let's divide it up before the guests arrive.

Living Will

Declaration made this _____day of ______, 199 ___.

I, __________________________, willfully and voluntarily make known my desire that my dying not be artificially prolonged under the circumstances set forth below, and I do hereby declare:

If at any time I have a terminal condition and if my attending or treating physician and another consulting physician have determined that there is no medical probability of my recovery from such condition, I direct that life-prolonging procedures be withheld or withdrawn when the application of such procedures would serve only to prolong artificially the process of dying, and that I be permitted to die naturally with only the administration of medication or the performance of any medical procedure deemed necessary to provide me with comfort care or to alleviate pain.

It is my intention that this declaration be honored by my family and physician as the final expression of my legal right to refuse medical or surgical treatment and to accept the consequences for such refusal.

In the event that I have been determined to be unable to provide express and informed consent regarding the withholding, withdrawal, or continuation of life-prolonging procedures, I wish to designate, as my surrogate to carry out the provisions of this declaration:

Name: ____

Address: ____

____

Phone: ____

I understand the full importance of this declaration, and I am emotionally and mentally competent to make this declaration.

____

Witnesses:

____________________	____________________
(Signature of Witness)	(Signature of Witness)
____________________	____________________
(Address)	(Address)
____________________	____________________
____________________	____________________
(Phone)	(Phone)

DURABLE POWER OF ATTORNEY

BY THIS DURABLE POWER OF ATTORNEY I,________ __________ of ________ County,__________, appoint as my attorney in fact to manage my affairs.

This durable power of attorney shall not be affected by any physical or mental disability that I may suffer except as provided by statute, and shall be exercisable from this date. All acts done by my attorney pursuant to this power shall bind me, my heirs, devisees and personal representatives. This power of attorney is nondelegable.

All of my property and interests in property are subject to this durable power of attorney.

Without limiting the broad powers conferred by the preceding provisions, I authorize my attorney in fact to:

1. Collect all sums of money and other property that may be payable or belonging to me, and to execute receipts, releases, cancellations or discharges.
2. Settle any account in which I have any interest and to pay or receive the balance of that account as the case may require.
3. Enter any safe-deposit box or other place of safe-keeping standing in my name alone or jointly with another and to remove the contents and to make additions, substitutions and replacements.
4. Borrow money on such terms and with such security as my attorney may think fit and to execute all notes, mortgages and other instruments that my attorney finds necessary or desirable.

5. Draw, accept, endorse or otherwise deal with any checks or other commercial or mercantile instruments, specifically including the right to make withdrawals from any savings, checking and/or money market account.
6. Redeem bonds issued by the United States government or any of its agencies, any other bonds and any certificates of deposit or other similar assets belonging to me.
7. Sell bonds, shares of stock, warrants, debentures or other assets belonging to me, and execute all assignments and other instruments necessary or proper for transferring them to the purchase or purchasers, and give good receipts and discharges for all money payable in respect to them.
8. Invest the proceeds of any redemptions or sales and any other of my money, in bonds, shares of stock and other securities as my attorney shall think fit.
9. Vote at all meetings of stockholders of any company and otherwise act as my proxy in respect of my shares of stock or other securities or investments that now or hereafter belong to me, and appoint substitutes or proxies with respect to any of those shares of stock.
10. Execute on my behalf any tax return and act for me in any examination, audit, hearing, conference or litigation relating to taxes, including authority to file and prosecute refund claims, and enter into any settlements.

11. Sell, rent, lease or mortgage for any term, or exchange, any real estate or interests therein for such considerations and upon such terms and conditions as my attorney may see fit, and execute, acknowledge and deliver all instruments conveying or encumbering title to property owned by me alone as well as any owned by me or another party, jointly.
12. Prosecute, defend and settle all actions or other legal proceedings touching my estate or any part of it or touching any matter in which I may be concerned in any way.
13. Purchase bonds issued by the United States that can be applied at face or maturity value on account of estate tax liabilities.
14. Make gifts of any real or personal property as my attorney shall see fit, not, however, exceeding $10,000 per year, per person.
15. Do anything regarding my estate, property and affairs that I could do myself.

The powers conferred upon my attorney in fact extend to all of my right, title and interest in property in which I may have an interest jointly with any other person, whether in an estate by the entirety, joint tenancy or tenancy in common.

This instrument is executed by me in the State of __________, but it is my intention that this power of attorney shall be exercisable in any other state or jurisdiction where I may have any property or interests in property.

I hereby confirm all acts of my attorney in fact pursuant to this power.

Any act that is done under this power between the revocation of this instrument and notice of that revocation to my attorney shall be valid unless the person claiming the benefit of the act had notice of that revocation.

IN WITNESS WHEREOF, I have set my hand on __________, 199 ____.

Signed and delivered in presence of:

____

____
Witness

____
Witness

STATE OF ____________________________________
COUNTY OF __________________________________

The foregoing instrument was acknowledged before me this _____day of____________,199____ by______________, who is personally known to me or has produced satisfactory identification.

____
Signature of Notary Public
Commission Information
and Stamp:

CHAPTER SEVEN

UNCOMMON PROBLEMS AND EASY EXPLANATIONS

I own a Cadillac. I intend to give it to my son when I die and I have made a provision for this gift in my will. What happens if I give it to him before my death?

The gift to your son in your will is going to be satisfied by "ademption."

What is "ademption by satisfaction"?

A gift by a will-maker during the will-maker's lifetime is treated as the satisfaction or completion of the same gift within a will.

My spouse and I own everything jointly. What happens if we both die at the same time?

In most states, a simultaneous death law addresses this situation. Under most of these laws, each of you will be deemed to own a one-half interest in all jointly owned assets. The assets will pass according to the terms of your two wills, probated separately, or distributed according to other state laws in the event you both die without wills. When signing wills, you can direct the sequence of your death, if it cannot be determined (out of wreckage, etc.), and the sequence that you select will be followed. I can't make any jokes about this kind of common tragedy.

If some dead guy leaves me something in his will, must I accept it?

No. You may file a "disclaimer."

What is a "disclaimer"?

A disclaimer is a signed document filed with the probate court, within a specified period of time that varies from state to state, by which a beneficiary rejects an inheritance. The inheritance passes as if the disclaiming person predeceased the dead guy.

My husband pays alimony to me. If he dies, can I continue to receive alimony from his assets?

Usually. In most states, alimony is a claim that can be made against a deceased ex-spouse's estate. You may wish to review the divorce decree or settlement agreement that was put into place at the time of your breakup. The decree or agreement should have been written to deal with this situation.

Do stepchildren receive inheritances the same way as blood offspring?

No. Usually stepchildren are excluded from the definition of children in most state laws and standardized forms of wills. If you wish to make a bequest or devise to a stepchild, you may do so by specific provisions in your will.

What happens if my attorney makes a mistake in my will or trust?

He or she may be subject to malpractice. Won't that make you happy?

In many cases, a will or trust can be corrected through "reformation."

What is "reformation"?

Reformation is the name for court proceedings by which a judge clarifies or corrects a will or trust in order to make it carry out the will-maker's or trust-maker's intentions.

If I become too feeble to make medical choices, can I name someone to make them for me?

Yes. Sign a "designation of health care surrogate."

What is a "designation of health care surrogate"?

It's a written document, like a power of attorney, that authorizes someone, such as your spouse or child, to deal with doctors when you can't. A sample follows this question-and-answer session. Think twice before naming Dr. Jack Kervorkian.

I love my body. I want to donate body parts to different people. How do I do this?

Sign a "uniform donor card."

What is a "uniform donor card"?

A uniform donor card is a written statement, usually signed and witnessed, by which a person designates body parts to be given to surviving individuals who need transplants or to institutions for medical study or transplant.

A form for donating body parts is located at the posterior of this chapter.

I thought at least he would survive the honeymoon. Maybe it was my cooking.

DESIGNATION OF HEALTH CARE SURROGATE

Name: ____

(Last) (First) (Middle Initial)

In the event that I have been determined to be incapacitated to provide informed consent for medical treatment and surgical and diagnostic procedures, I wish to designate as my surrogate for health care decisions:

Name: ____

Address: ____

Phone: ____

I fully understand that this designation will permit my designee to make health care decisions and to provide, withhold or withdraw consent on my behalf; to apply for public benefits to defray the cost of health care; and to authorize my admission to or transfer from a health care facility.

Signed: ____

Date: ______________________________ , 199 __________

Witnesses: 1. __________________ 2. __________________

UNIFORM DONOR CARD

The undersigned hereby makes this anatomical gift, if medically acceptable, to take effect on death. The words and marks below indicate my desires:

I give:

(a) ___ any needed organs or parts;

(b) ___ only the following organs or parts:

____

____

for the purpose of transplantation, therapy, medical research or education;

(c) ___ my body for anatomical study if needed.

Limitations or special wishes, if any: ______________________

(If applicable, list specific donee)

Signed by the Donor and the following Witnesses in the presence of each other:

______________________	______________________
(Signature of Donor)	(Date of birth of Donor)
______________________	______________________
(Date signed)	(City and State)
______________________	______________________
(Witness)	(Witness)
______________________	______________________
(Address)	(Address)

CHAPTER EIGHT

I.R.S. AGENTS AND COUNTER-ESPIONAGE TO SAVE TAXES

How can I save estate taxes?

Attempt to overthrow the I.R.S. I personally oppose this civil disobedience. If any institution in our present world civilization is immortal, it's the I.R.S. The best you can do is eliminate or minimize your payment to them. I.R.S. agents should be thought of as poor relations who yearn to receive some inheritance.

I love the I.R.S. and I'm going to keep loving them until after my own estate tax proceedings are over and I've been dead for at least five years.

What is the most surefire way to eliminate or minimize estate taxes?

The first most certain way is to reduce the size of your estate to under the amount of $625,000. Estates worth $625,000 or less are not subject to any federal estate taxes. Under the new 1997 tax law, this amount gradually increases to $1,000,000 by the year 2006. The second most certain way is to give everything to your spouse. The I.R.S. assesses absolutely no tax on gifts or legacies between married people.

New York assesses taxes on assets over $115,000 (until the year 2000); the I.R.S. is the friendly adversary for purposes of this discussion, not the state.

How can I reduce the size of my estate?

By making gifts during your lifetime.

However, one's financial security through old age is more important than victory over the I.R.S. by descendants.

Should I die broke?

Ideally, to save estate taxes, you should die broke. This, of course, is extreme, usually impractical and flat-out offensive to most of the living and the dead alike. (Some people, of course, give everything to charity and thereby cheat Uncle Sam.)

By the way, Chief Justice Fred Vinson of the United States Supreme Court died broke.

If you die broke, you will be a dead deadbeat. I must say that is an interesting aspiration.

If I am financially secure and wish to reduce my estate through gifts, what gifts should I make?

You are entitled to an annual gift tax exclusion. In addition, you are entitled to a "unified credit."

What is the "annual gift tax exclusion"?

Anybody can give anybody else $10,000 each year without incurring any gift taxes. This $10,000 amount increases after 1998 for the cost of living.

Can I make these $10,000 gifts only to my family? What about friends?

You can make them to any individual, family, friend or foe.

What is a "unified credit"?

Everyone also is entitled to a unified credit by which

$600,000 can be given, through lifetime gifts or under one's will, to anyone. This money passes free of gift taxes and estate taxes. The $600,000 increases to $625,000 in 1998 and gradually to $1,000,000 by 2006.

If my spouse and I each have $600,000, can we each use our unified credits to give a total of $1,200,000 to our children?

Yes. Each of you can set up a certain trust, usually called a "by-pass trust" or "credit-shelter trust," to accomplish this goal.

How do I set up a by-pass trust or credit-shelter trust?

This is how it works. You each create a trust funded with $625,000. The trust can be lifetime or testamentary. Income, and principal with certain restrictions, can be paid to the surviving spouse for life and then, upon the surviving spouse's death, "his" and "her" unified credits, the two of them totaling $1,200,000, will pass tax-free to the children.

What gift or estate taxes do I pay for tranfers to my spouse?

None, because of the "unlimited marital deduction."

What is the "unlimited marital deduction"?

A law that enables spouses to give unlimited amounts of money to each other without paying any gift or death taxes.

I wish to give all of my money to my wife but make sure it passes to my children upon her death. My wife is such a bombshell that she inevitably will remarry; I don't want her next husband to receive my hard-earned money. What do I do?

Persuade your wife to undergo plastic surgery for ugliness.

A "Q-TIP" trust is the thing for you.

Did you say "Q-TIP"? "Q-TIP" trust? What's that? I couldn't hear you.

I repeat: A Q-TIP trust is the thing for you. It has nothing to do with dirty ears, but you may wish to check your hearing. Q-TIP is an acronym from the I.R.S. that stands for "qualified terminable interest property." If you put your assets into a Q-TIP trust, your wife can receive its income, and principal, during her lifetime. Upon her death, assets can pass as you direct. No taxes are due on your death; taxes may be due on her death.

I wish to leave money in a marital trust but without the restrictions of the Q-TIP trust. Can you suggest an alternative?

Sure, especially since you ask so nicely.

You can establish a more flexible trust — with income and principal distributable to your wife as you like — and give her a power of appointment to be exercised in her will. The principal will be distributed according to her wishes, when she keels over, and her wishes may match your own.

If I make legacies under my will to charities, such as my church or temple, will this save estate taxes?

Yes, if the charity is a qualified nonprofit organization. Charitable contributions can mean big tax savings, but they may be made at the expense of family members. There is a trade-off between paying money to the I.R.S. and giving money to charity.

Check out your charities. The I.R.S. doesn't recognize the "Church Cable Company" as not-for-profit. Be careful of temples of doom.

I understand from my life insurance agent that life insurance can eliminate estate taxes. True or false?

Life insurance policies and proceeds, in and of themselves, do not pass free of death taxes. Life insurance is not exempt from estate taxes the way tax-exempt bonds may be exempt from income taxes. Life insurance trusts, however, may reduce or even eliminate death taxes.

What is a life insurance trust? Are there benefits to setting one up?

If you (a) set up an irrevocable trust for the benefit of others during your lifetime, (b) put money into it to enable the Trustee to purchase insurance on your life and (c) die, the life insurance proceeds payable to the trust and its beneficiaries should not be subject to estate taxes. The trust should be excluded from your estate. All assets in your estate remain taxable.

If you create an irrevocable trust for the benefit of others and put money into it, for the purchase of life insurance or other assets such as stocks and bonds, you essentially are reducing the size of your estate to save death taxes. The choice of life insurance as an asset of the trust is an investment selection that has advantages and disadvantages when compared with other types of investments.

I understand you can place your house into a trust in order to save estate taxes. Is this true?

Yes. You are referring to "qualified personal residence trusts." These also are known as "house GRITS"; GRIT is an acronym for Grantor retained income trust. It is not a good ol' home filled with Southern, cooked cornmeal.

How do I set up a "house GRIT"?

A short how-to on house GRITS. Joe Blow, Sr., deeds his home into a trust for a period of time not exceeding twenty years. Joe, Sr., retains the use of the house during the selected time period. After the time expires, it passes to his son, Joe Blow, Jr. At the time the trust is established and the deed is signed, Joe, Sr., makes a present gift of a future interest to Joe, Jr. Gift taxes on the value of the future interest may be owed, but the $10,000 annual gift tax exclusion (increasing gradually to $1,000,000 by 2006) and $625,000 unified credit exemption (or more if adjusted for the cost of living after 1998) may be available. If the house is worth $250,000 now, it may be worth $500,000 when it passes to Joe, Jr., because of appreciation. Death taxes on the appreciated $250,000 can be saved.

Qualified personal residence trusts or house GRITS are complicated. Don't try them at home. (Get it? "At home"?)

Are there any other tax-saving techniques out there?

Yes. For examples, charitable trusts and foundations, discussed later, and family asset freezes, discussed next. These are highly sophisticated techniques. If you have a lot of money, discuss all possibilities with your lawyer, accountant and most trusted friend.

How do you accomplish a family asset freeze?

You and your family members each sit on a block of ice until you freeze your assets off.

You put assets into a corporation, partnership or other entity and make gifts to your family members of fractional interests. The interests you give away reduce the size of your estate. The interests you keep will become part of your taxable estate, but the I.R.S. may recognize favorably low valuations of these interests because of their lack of marketability (i.e., the interests are hard to sell to third-party buyers). If you only keep a minority stake in a family company, your estate's value may benefit from a minority discount (i.e., the lack of control by a minority). Minority discounts are available to White Anglo-Saxon Protestants.

If I'm dead, buried and six feet under, how will the I.R.S. collect taxes from me?

They collect from the assets you leave behind. Shrouds have no pockets. Even if you are buried with your gold, I.R.S. agents will exhume.

If I make gifts to my grandchildren instead of my kids, will that save death taxes?

It will save on the death taxes payable by your children. Beware, however, of "generation-skipping taxes."

What are "generation-skipping taxes"?

Special taxes assessed on gifts, lifetime or testamentary, to younger generations that involve skipping older generations. These taxes are as high as estate taxes, but everyone gets a $1,000,000 generation-skipping tax-exemption.

What is the $1,000,000 generation-skipping tax-exemption?

Everyone can give away $1,000,000 (assuming they have it!) to grandchildren, great-grandchildren and the like without having to pay the high generation-skipping taxes. This $1,000,000 increases for cost of living after 1998.

Is a tax return filed after I've cashed in my chips for good?

Yes. It's called an "estate tax return."

What's involved in an estate tax return?

All of the dead person's assets are appraised and, after the unified credit ($625,000 in 1998 and increasing to $1,000,000 gradually by 2006), administration expenses and other items, taxes are computed in the return based on the value of the estate.

Who files the estate tax return?

Your Executor, usually, or the Trustee of your living trust.

How are assets in the estate tax return valued?

Real estate, businesses, art and other assets are appraised by experts. The experts use professional appraisal methods. The valuation is as of the date of death or six months from the date of death; the I.R.S. gives your Fiduciary a choice of these two dates and your Fiduciary should pick the most favorable date for reducing taxes.

I bought IBM at $25 per share decades ago and I will never sell it. If I did, I would pay big capital gains taxes. Will my estate pay big capital gains taxes if this IBM stock must be sold to pay death taxes?

No. Your estate receives a "step-up basis."

What is a "step-up basis"?

The value of the IBM shares, for purposes of estate taxes, is the value as of your date of death.

Step-up basis is also traveling from 0 to 60 miles per hour in 6 seconds.

If I give this IBM to my children now, will they receive a step-up basis?

No. They will have to use your cost basis as their cost basis.

I have $100,000 that I own jointly with my husband. We want to avoid probate, leave everything to each other and ultimately to our children. What's the plan for us?

First, make sure all of your assets are owned jointly by you and your hubby with rights of survivorship. This will insure that the survivor of you two takes all without probate. Second, after one of you is dead, the survivor should put all assets into joint names, again with rights of survivorship, with your children, so they will receive everything automatically when the survivor dies.

I am a sophisticated person. What are some super-duper sophisticated estate-planning techniques?

"GRATs" and "GRUTs."

Please stop grunting. What do you mean by GRATs and GRUTs?

GRAT is an acronym for "Grantor retained annuity trust."

In a GRAT, a trust-maker sets up a trust under which he retains the right to receive a fixed annual dollar amount, an annuity, for a set number of years. The annuity is usually a percentage, say 5 percent, of the initial value of the property put into the trust. After the specified number of years passes, the property held in the GRAT passes to beneficiaries designated by the trust-maker. The assets may pass to a spouse, children, friends or anyone at all. When these beneficiaries receive the property, it's likely to have appreciated in value since the time when the trust-maker created the trust. All of this appreciation passes to the beneficiaries free of any federal gift or estate tax.

A GRUT is a "Grantor retained unitrust." A GRUT is the same thing as a GRAT, except that the annuity isn't fixed on the basis of the initial value of the trust's assets but is variable depending upon the value of assets each year.

My family owns an historic mansion on two hundred acres of land. It is our main asset. When my spouse and I are both dead, we are concerned that our children will be forced to sell it in order to pay estate taxes. How can we avoid this loss of real estate?

I suggest that you consider a "preservation easement." You must consult a specialist on this subject.

What is a "preservation easement"?

A deed by which an individual makes a gift of the exterior facades of a mansion or the development rights on vacant land to a qualified organization, such as the National Trust for Historic Preservation. The gift is irrevocable and, therefore, lasts forever.

How do the tax benefits work? What happens to the property?

When the gift of the preservation easement is made, you receive a charitable deduction. Neither you nor your spouse nor your children, grandchildren or great-grandchildren will be allowed to change the exterior of the mansion or build on the vacant land without the consent of the charity/owner of the preservation easement. The value of the mansion and the land is reduced drastically in the valuation of assets for the computation of estate taxes. The savings can be enormous and enable your family to "save the farm," or "save the ranch," or save some real estate.

The deceased left all of you equal shares.

TAX FREE ESTATES

THE UNIFIED CREDIT'S ESCAPE FROM TAX BITES

YEAR	VALUE OF ASSETS
1997	$ 600,000
1998	$ 625,000
1999	$ 650,000
2000	$ 675,000
2001	$ 700,000
2002	$ 700,000
2003	$ 700,000
2004	$ 850,000
2005	$ 900,000
2006	$ 1,000,000

Living Trust With Tax-Saving Provisions

199___ REVOCABLE TRUST,

between

____

as Donor,

and

____

as Trustee

Dated: ______________________, 199___

______________________ 199__ REVOCABLE TRUST AGREEMENT dated ________________, 199___, between ____________________________, as Donor, and__________ ___________________, as Trustee.

1. **TRUST PROPERTY.** The Donor gives the property described in Schedule A (attached) to the Trustee as the principal of the Trust. The Trustee acknowledges the receipt of such property and agrees to manage and dispose of it, together with any additions and income, according to this Agreement.

2. **LIFETIME MANAGEMENT AND DISPOSITIONS.** During the Donor's lifetime, the Trustee shall (a) invest the principal of the Trust as the Donor may direct from time to time by a written instrument signed by the Donor and delivered to the Trustee and (b) pay to or apply for the benefit of the Donor as much of the net income and principal of the Trust, both as the Donor shall direct by written instructions signed and delivered to the Trustee.

3. **DISPOSITIONS AT DEATH.** Upon the Donor's death, the principal and interest of the Trust shall be distributed as follows:

A. Credit Shelter Trust:

(1) If the Donor's spouse, _________ ________________, shall survive the Donor, the Trustee shall set aside in a separate trust an amount equal to the largest sum that can pass free of federal estate tax by reason of the unified credit and the state death tax credit allowable to the Donor's estate, reduced by (i) the value of all property, including gift taxes, included in determining the value of the Donor's gross estate for federal estate tax purposes which does not qualify for the marital or charitable deduction, (ii) the value of all of the Donor's adjusted taxable gifts, and (iii) the amount of all charges to principal that are not allowed as

deductions in computing the federal estate tax on the Donor's estate, other than income taxes, and further reduced to the extent necessary so that state death taxes will not be imposed or increased with respect to the Donor's estate solely by reason of this provision. The computation of this sum shall be controlled by the valuations and other determinations in the federal estate tax proceedings for the Donor's estate.

(2) The net income of this Trust shall be paid to or applied for the benefit of the Donor's spouse, in convenient installments but no less often than annually.

(3) The Trustee is authorized to pay to or apply for the benefit of the Donor's spouse such part or all of the principal of this Trust as the Trustee shall deem advisable in its absolute discretion for any reason whatsoever, even though any such payment or payments shall result in the termination of the trust.

(4) Upon the death of the Donor's spouse, the principal of this Trust remaining at that time shall be disposed of in accordance with the provisions of Subdivision C, applied as if the Donor had survived the Donor's spouse.

SELECT AN ALTERNATIVE, CHOICE I or II:

CHOICE I:

OUTRIGHT DISPOSITION OF BALANCE OF TRUST TO SPOUSE

B. Disposition of Balance to Spouse.

If the Donor's spouse, _______________, shall survive the Donor, the principal of the Trust remaining at that time (after the dispositions in Subdivision A of this Section) shall be distributed to such spouse outright.

CHOICE II:

DISPOSITION OF BALANCE OF TRUST TO SPOUSE IN FURTHER QTIP TRUST

B. QTIP Trust.

(1) If the Donor's spouse, _______________, shall survive the Donor, the principal of the Trust remaining at that time (after the dispositions in Subdivision A of this Section) shall be held by the Trustee in a separate Trust for the benefit of the Donor's spouse with the income and principal thereof to be disposed of as follows:

(a) The net income of this Trust shall be paid to the Donor's spouse, in convenient installments but no less often than annually.

(b) The Trustee is authorized, at any time and from time to time, to pay to the Donor's spouse such part or all of the principal of this Trust as the Trustee shall deem advisable in his absolute discretion for any reason whatsoever, even though any such payment or payments shall result in the termination of the Trust.

(c) Upon the death of the Donor's spouse, the principal of this Trust at that time remaining shall be disposed of in accordance with the provisions of Subdivision C, applied as if the Donor had survived the Donor's spouse.

(2) By the disposition for the benefit of the Donor's spouse under the foregoing provisions of this Subdivision, the Donor intends to obtain for his estate the right to elect the benefit of the marital deduction provided for in the Internal Revenue Code, as amended from time to time (the "Code").

(3) The Donor directs that this Subdivision and

any other provision of this Agreement shall be construed and interpreted and, if necessary, modified and limited, so as to permit the disposition for the benefit of the Donor's spouse to qualify for the marital deduction: (i) the term "net income" as used in paragraph (1)(a) of this Subdivision shall mean an amount not less than the income to which the Donor's spouse must be entitled in order to obtain for the Donor's estate the benefit of the marital deduction with respect to such Trust; (ii) such Trust shall not be funded with any assets in the Donor's gross estate, or their proceeds, with respect to which marital deduction would not be allowed if such assets passed from the Donor to the Donor's spouse; and (iii) such Trust shall not be funded with any assets, or their proceeds, which are subject to any estate, transfer, succession, inheritance or other death duty or tax by any foreign country or subdivision thereof, except to the extent that such assets or proceeds are needed to satisfy such legacy.

(4) The Donor authorizes the Trustee to elect (or not to elect) to treat all or any portion of property passing for the benefit of the Donor's spouse under this Subdivision B or otherwise, as so-called qualified terminable interest property as defined in Section 2056(b)(7) of the Code, so that such property, or such specific portion thereof, shall qualify for the federal estate tax marital deduction. The determination of the Trustee whether or not to make such election, and the nature and extent of the property to which any such election relates, shall be binding upon all interested parties. The Trustee shall have complete discretion in determining whether, and the extent to which, to make any such election.

SELECT AN ALTERNATIVE, CHOICE III, IV OR V:

CHOICE III

OUTRIGHT EQUAL DISTRIBUTIONS TO CHILDREN

C. If the Donor's spouse shall not survive the Donor, the principal of the Trust remaining at that time shall be distributed to the issue *per stirpes* of the Donor who shall be living at that time.

CHOICE IV

EQUAL DISTRIBUTIONS TO CHILDREN, PARTLY IN TRUST:

C. If the Donor's spouse shall not survive the Donor, the principal of the Trust remaining at that time shall be divided and set aside in equal shares, one share for each child of the Donor who shall be living at that time, and one share for the issue collectively who shall be living at that time of each child of the Donor who shall not be living at that time.

(1) Each such share so set aside for ______________ shall be distributed to such child outright.

(2) Each such share so set aside for ______________ shall be distributed to the Trustee of the __________199___ Trust and held in accordance with the terms of such Trust.

(3) Each such share so set aside for the issue collectively of a deceased child of the Donor shall be distributed to such issue *per stirpes.*

CHOICE V

UNEQUAL DISTRIBUTIONS TO CHILDREN, PARTLY IN TRUST:

C. If the Donor's spouse shall not survive the Donor, the principal of the Trust at that time remaining shall be distributed as follows:

(1) ________ % shall be distributed outright to the

Donor's child, ____________________, or if such child shall not be living at that time, to the issue of such child as shall be living at that time *per stirpes.*

(2) _______% shall be distributed outright to the Donor's child, ______________________, or if such child shall not be living at that time, to the issue of such child as shall be living at that time *per stirpes.*

(3) _______% shall be distributed to the Trustee of the ________________ 199___ Trust for the benefit of the Donor's child, ______________________, or if such child shall not be living at that time, to the issue of such child as shall be living at that time *per stirpes.*

(4) If any of subparagraphs (1), (2) or (3) of this Subdivision shall fail to take effect, the property to be distributed or set aside under such subparagraph shall be distributed proportionately in accordance with such of the provisions of this Subdivision as shall take effect.

4. **TRUSTEE PROVISIONS.**

A. Governing Law and Powers of Trustee. This Trust shall be governed by the laws of the State of ______________. The Trustee shall have all powers provided under the laws of the State of ____________________.

B. Successor Trustee. If for any reason _______ ____________________ is unable to serve as Trustee, ________________ shall serve as successor Trustee.

C. Waivers. Any law or local court rule now or hereafter in effect requiring qualification, administration or accounting by the Trustee to any court shall not apply to the Trustee and are hereby waived by the Donor.

5. **REVOCABILITY.** The Donor shall have the right to amend or revoke this Agreement at any time and from time to time by a written instrument, signed, acknowledged

and delivered to the Trustee, provided that the Trustee's duties may not be increased nor its compensation decreased without its consent.

6. **PAYMENT OF DEBTS, EXPENSES AND TAXES.** Following the death of the Donor and upon the request of the legal representatives of the Donor's estate, the Trustee shall pay to such legal representatives or directly to the appropriate person or entities from the principal of the trust, such sum or sums as shall be certified to them by such representatives to be due for any part or all of (a) the debts of the Donor, (b) the expenses of the Donor's last illness and funeral expenses, (c) the expenses of administering the estate of the Donor, and (d) any estate, transfer or other inheritance taxes, including any interest or penalty thereon imposed by the United States or any state thereof or any foreign country or subdivision thereof as a result of the death of the Donor, without regard to the amount, if any, of any such taxes, including interest and penalties, which would otherwise be payable from the principal of the trust under any applicable provision of law.

IN WITNESS WHEREOF, the Donor and Trustee have executed this Agreement as of the date shown above.

______________________	______________________
Witness	, Donor
______________________	
Witness	
______________________	______________________
Witness	, Trustee
______________________	
Witness	

STATE OF ______________________________
COUNTY OF ____________________________

We ________________ , ____________________ and ____________________ , the Donor, Trustee and the Witnesses, respectively, whose names are signed to the attached instrument, were sworn, and declared to the undersigned officer that the Donor in the presence of the Witnesses signed such instrument as the ______________ 199___ Revocable Trust, that the Donor signed, and that each of the Witnesses, in the presence of the Donor and in the presence of each other, signed the Trust as a Witness.

____
Donor and Trustee

____
Witness

____
Witness

Subscribed and sworn to before me by____________ ____________, the Donor and ________________the Trustee, and by________________and ______________ , the Witnesses, on ______________, 199___, all of whom personally appeared before me. ______________ is personally known to me or has produced satisfactory proof of identification. __________________ is personally known to me or has produced satisfactory proof of identification. __________________ is personally known to me or has produced satisfactory proof of identification.

____
Signature of Notary Public
Commission Information and Stamp:

SCHEDULE A

[List of Assets; for example, "all of the Donor's assets, including, without limitation, all stocks, bonds, checking and savings accounts, real estate, partnership interests and tangible personal property]

CHAPTER NINE

MARRIAGE CONTRACTS

What are "premarital contracts" or "antenuptial agreements"?

Contracts between a man and a woman in contemplation of marriage in which both agree to settlement terms if they divorce, inheritance rights upon death and any number of other aspects of their relationship during and after matrimony.

What do marriage contracts have to do with wills?

A lot.

Marriage contracts can create, modify or even eliminate inheritances.

In writing wills, one must always consider the existence of any premarital contracts. Antenuptial agreements, in and of themselves, involve family planning and the transfer of assets.

Aren't antenuptial contracts antinuptial?

Surely these agreements are not romantic. An enamored suitor hardly would propose, "Will you marry me, my true love...but first sign a premarital contract prepared by my attorneys, Wham, Bam & Slam?" It is hard to imagine a cinema love scene in which Tom Cruise and Michelle Pfeiffer are kissing in a conference room filled with two sets of attorneys. So much for Jackson Browne's song "Lawyers in Love."

A relationship may be doomed if either lover demands a premarital agreement. To even ask for one may be interpreted to imply a lack of love. Even if the man and woman agree that they should determine in advance the economic consequences of marriage, but disagree on particular terms, they may cancel their wedding. If the man and woman do sign a contract and marry, one or the other may regard certain terms as unfair and the existence of the agreement can lead to resentment, argument and divorce. To continue this woeful line of thought, contested divorces frequently involve attacks on the validity of the prenuptial agreements themselves. It seems that if a romance at any stage has underlying problems, a premarital contract can add to them.

What are the advantages of a premarital contract?

First, privacy. People who cherish privacy appreciate the fact that antenuptial contracts enable them to resolve marital rights with their prospective spouses in a private conference room rather than a public courthouse under harsh media lights.

Second, choice of law. Parties can pick the marital law to apply to them if they divorce. They can make certain choices otherwise decided by state legislatures and judges.

Third, the protection of children. A person with children by a prior spouse may sign an antenuptial agreement with a second spouse to preserve his or her children's inheritances. Anyone who has ever read *Cinderella* knows how ugly relationships between stepparents and stepchildren can be.

Fourth, the preservation of family fortunes. To protect a family fortune, a person, or even parents or grandparents, may demand a premarriage contract to prevent family assets from going to an in-law.

Fifth, disclosure and openness. The disclosure of assets usually required for a valid premarital contract may raise or lower expectations about lifestyles but, in any event, stop misconceptions. Revelations about character that often occur in negotiations may be more significant than the disclosure of assets. A man and woman may discover if their family philosophies and ideas about fairness and sharing are compatible. It is often easier to reach understandings about marital rights during the spirit of compromise characterizing courtship than afterward, when love may change to hate. Benjamin Franklin wrote in *Poor Richard's Almanac,* "Keep your eyes wide open before marriage, half-shut afterwards."

What are the disadvantages of antenuptial agreements?

First, divisiveness. If a husband and wife sign a premarital agreement, they may not act as a team. They may not carry the same financial interests "during good times and bad times."

Second, psychological problems. Some people will negotiate antenuptial agreements for psychological security. Their property may be more important to them than their personal relationships, especially if they have worked hard for their money and care more about keeping their stash than their spouse. Psychological problems may be hardened in the concrete of these contracts.

Third, control. Some individuals use antenuptial

contracts to dominate their spouses through economic control.

Are antenuptial agreements new?

No. In the early 1800s, a couple named Andrews signed such a contract and it was enforced. Divorce goes way back. The war between the sexes is the longest war of all. When will it be peacetime?

What are some of the subjects that antenuptial agreements cover?

Here are a few examples.

A contract may provide that one spouse pay the other a certain amount of money in alimony each year in the event they are divorced.

An agreement may provide that the couple will have sexual relations five times each week. Pant, pant, pant.

These agreements may address child rearing. A bride and groom, looking to the future, may decide that they want their children raised in accordance with the principles of a certain religion.

Between two wage-earning career people, marriage is just like a joint venture and the joint venturers both may wish to contribute income and settle in writing how money will be invested and how money will be spent.

Do most lawyers recommend premarital contracts?

I don't know. It's privileged communication between lawyers and clients. I can tell you this:

In 1900, the ratio between marriages and divorces was roughly 13 to 1. Nowadays, about one out of every